CELL BIOLOGY RESEARCH PROGRESS

LEYDIG CELLS

STRUCTURE, FUNCTIONS AND CLINICAL ASPECTS

Cell Biology Research Progress

Additional books and e-books in this series can be found on Nova's website under the Series tab.

CELL BIOLOGY RESEARCH PROGRESS

LEYDIG CELLS

STRUCTURE, FUNCTIONS AND CLINICAL ASPECTS

BRUNO SOLOMON
EDITOR

NOTICE TO THE READER

Library of Congress Cataloging-in-Publication Data

ISBN: 978-1-53617-282-9

Published by Nova Science Publishers, Inc. † New York

CONTENTS

PREFACE

The main function of the Leydig cells is to synthesize testosterone, but growing evidence supports the idea that Leydig cells are important in regulating the testicular immune environment. As such, Leydig Cells: Structure, Functions and Clinical Aspects discusses Leydig cells serving as immunological sentinels in the testis.

Next, the authors explore the commercially available immortalized Leydig cell lines (TM3, R2C, MA-10 and MLTC-1) derived from rat or mouse and the results of studies using these cells.

Additionally, the effects of the most commonly used environmental contaminants on Leydig cell function are discussed in the context of the impaired of steroidogenic pathway.

The results of several studies are presented which demonstrate that lycopene supplementation improves and maintaines the normal level of testosterone by improving the expression of StAR protein and steroidogenic enzymes in the Leydig cells of polychlorinated biphenyl exposed rats.

Chapter 1 - The testis is a remarkable immunoprivileged organ, as it possesses a distinct immune environment that protects male germ cells from a detrimental immune response. It is also well equipped with innate defense mechanisms that defend it against microbial infection. Testicular cells play critical roles in maintaining the special immune environment.

The main function of the Leydig cells is the synthesis of testosterone, but growing evidence supports the idea that Leydig cells are important in regulating the testicular immune environment. In addition to testosterone synthesis, Leydig cells produce various immunoregulatory factors that favor the immunoprivileged status of the testis. In addition, Leydig cells abundantly express various pattern recognition receptors that initiate innate cellular immune responses against microbial infections. This chapter discusses Leydig cells serving as immunological sentinels in the testis.

Chapter 2 - In recent years, the incidence of male infertility is considered as a global health problem. Testicular damage and functional changes in the reproductive system is great concern for male reproduction. Testes contain several cells types such as Sertoli, Leydig and spermatogenic cells. Leydig cells which of them located in interstitial compartment of testes play an essential role in maintaining spermatogenesis and biosynthesis of testosterone. In the male reproductive system, it is necessary to examine the steroidogenic pathway, through the expression levels of the genes and proteins involved, secondary messengers and many other factors responsible for cell function in order to reveal the conditions affecting androgen production. Today, cell culture studies are very important for animal welfare and provide scientists with the opportunity to investigate the molecular mechanism of diseases, to elucidate cell toxicity and to model models for various clinical problems. This chapter focuses on the comparison of commercially available immortalized Leydig cell lines derived from rat or mouse and the results of studies with these cells.

Chapter 3 - In the world, population growth, dietary habits, industrialization and the progress of agriculture have increased the frequency of exposure of people to various pollutants. Recent studies have shown that environmental pollutants cause many damages on male reproductive health. Environmental pollutants lead to male infertility, including the development of testicular cancer, decreased sperm count, impaired semen quality, deteriorated spermatogenesis and decreased testosterone production in the male reproductive system. It has been found that many environmental chemicals target endocrine system parameters that play a role in the reproductive function and cause various reproductive abnormalities. However, the basic mechanisms and molecular pathways involved in these anomalies have not been fully elucidated. In this context, it is important to detect changes in cell function of Leydig cells responsible for testosterone biosynthesis, which play a primary role in male reproductive system. This review discusses the effects of the most commonly used environmental contaminants on Leydig cell function through the impaired of steroidogenic pathway.

Chapter 4 - Leydig cells are present in the interstitial compartment of the mammalian testis and their main function is to produce testosterone which is essential for spermatogenesis and development of secondary sexual characters. LH is the primary regulator of Leydig cell function. Endocrine disruptors such as polychlorinated biphenyls, diethyl hexyl phthalate which are environmental contaminants cause adverse effects on Leydig cell structure and function hypertrophy and reduced capacity to produce testosterone. The authors' earlier studies proved that these disrupters altered the hypothalamic pituitary testicular axis in adult rats. The depletion of Leydig cellular antioxidant enzymes and increase in the levels of reactive oxygen species (ROS) and lipid peroxidation were observed in PCB exposed adult rats. The authors' studies demonstrated that PCB inhibits testosterone biosynthesis, Leydig cellular LH receptors, steroidogenic enzymes and antioxidant enzymes in adult rats. Lactational exposure of PCBs downregulated critical genes in Leydig cells of F1 male progeny. *In utero* exposure of phthalate downregulates critical genes in Leydig cells of F1 male progeny. Gestational exposure of phthalate induced hypermethylation in SP1 and Sp1 promoter in Leydig cells of F1 offspring rats.

Lycopene is a highly efficient antioxidant and has a singlet-oxygen and free radical scavenging capacity. It has protective effects against testicular toxicity, spermiotoxicity, cardiotoxicity and nephrotoxicity. The authors' studies demonstrated that lycopene supplementation improved and maintained the normal level of testosterone via improving the expression of StAR protein and steroidogenic enzymes in Leydig cells of PCB exposed rats. Ameliorative effect of α-tocopherol on PCB induced testicular Sertoli cell dysfunction in F1 prepubertal rats was also studied. Extensive studies on Leydig cells are in progress.

In: Leydig Cells
Editor: Bruno Solomon
ISBN: 978-1-53617-282-9

Chapter 1

IMMUNOLOGICAL FUNCTION OF LEYDIG CELLS

Ran Chen, Fei Wang and Daishu Han*
Institute of Basic Medical Sciences,
Chinese Academy of Medical Sciences, School of Basic Medicine,
Peking Union Medical College, Beijing, China

ABSTRACT

The testis is a remarkable immunoprivileged organ, as it possesses a distinct immune environment that protects male germ cells from a detrimental immune response. It is also well equipped with innate defense mechanisms that defend it against microbial infection. Testicular cells play critical roles in maintaining the special immune environment. The main function of the Leydig cells is the synthesis of testosterone, but growing evidence supports the idea that Leydig cells are important in regulating the testicular immune environment. In addition to testosterone synthesis, Leydig cells produce various immunoregulatory factors that favor the immunoprivileged status of the testis. In addition, Leydig cells abundantly express various pattern recognition receptors that initiate

* Corresponding Author's Email: dshan@ibms.pumc.edu.cn.

innate cellular immune responses against microbial infections. This chapter discusses Leydig cells serving as immunological sentinels in the testis.

Keywords: Leydig cell, testis, immune privilege, innate immune response, pattern recognition receptor

1. Histological Micro-Environment in the Testis

The adult mammalian testis has a high organized micro-environment considering its histological structure and cell types (Figure 1). The testis consists of two relatively separated regions: the seminiferous tubules and the interstitial spaces among the tubules. The seminiferous tubules are surrounded by peritubular myoid cells that form the wall of the tubules. Columnar Sertoli cells extend from the tubular wall to lumen and embrace germ cells to form the seminiferous epithelium where the spermatogenesis takes place. Leydig cells represent the majority of the interstitial cell populations and are the only cell type producing testosterone necessary for spermatogenesis. In addition to Leydig cells, blood vessels and various types of immune cells have been found in the interstitial spaces. Macrophages are a major type of immune cells. Other minor immune cells, including dentritic cells, lymphocytes, and mast cells, also reside in the testicular interstitial spaces. While testicular macrophages are initially believed to be the first line of defense against microbes invading from blood circulation, subsequent studies have indicated that the testicular macrophages predominantly display immunosuppressive properties (Bhushan and Meinhardt 2017). Other immune cells favor the immunoprivileged status in the testis. Moreover, testicular tissue-specific cells, including Leydig, Sertoli, and germ cells, are involved in the regulation of the immune micro-environment of the testis and the defense against microbial infection.

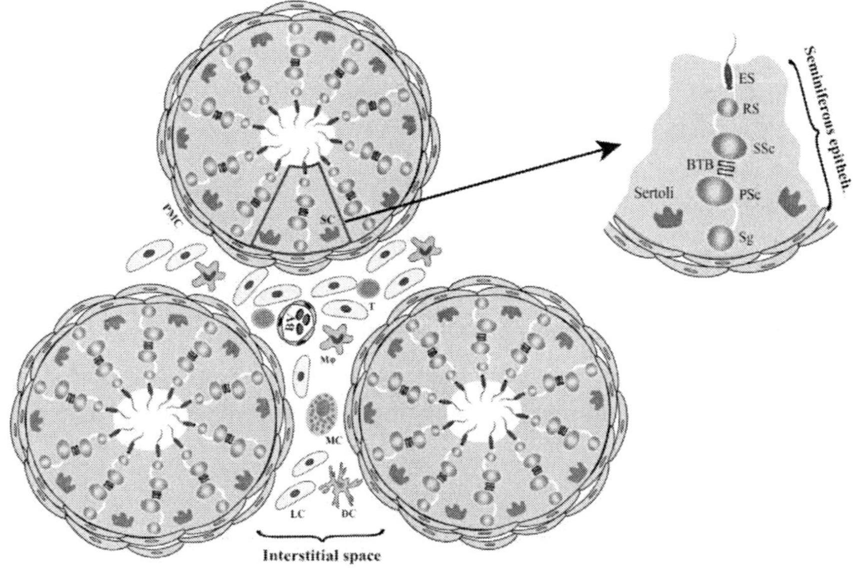

Figure 1. Histological schematic of the human testis. The testis is composed of two compartments: seminiferous tubules (ST) and interstitial space. The ST are surrounded by peritubular myoid cells (PMC) and Sertoli cells (SC) embracing developing germ cells, including spermatogonia (Sg), primary spermatocytes (PSC), secondary spermatocytes (SSC), round spermatids (RS), and elongated spermatids (ES). The interstitial space is composed of mainly Leydig cells (LC) and macrophages (Mφ), as well as minor T lymphocytes (T), dendritic cells (DC) and mast cells (MC). The blood vessels (BV) reside in the interstitial space.

2. Testicular Immune Environment

The mammalian testis owns a special immune environment considering its immune privilege and innate immunity, which are important in protecting immunogenic male germ cells from an adverse immune response and counteracting microbial infections (Zhao et al. 2014). The breakdown of the immune homeostasis in the testis may result in testicular dysfunction, thus leading to male subfertility or infertility. The mechanisms underlying testicular immune privilege have been intensively investigated for more than half a century. Both systemic immune tolerance and a local immunosuppressive environment are involved in the maintenance of the testicular immunoprivileged status. The local immunosuppression is conferred by multiple mechanisms, including tissue structure, cellular interaction, and immunosuppressive factors (Li, Wang, and Han 2012). The local innate defense against microbial infections has been revealed in the last decade. The mechanisms by which the testis defends microbial infections are mainly attributable to the pattern recognition receptors (PRRs)-initiated innate immune responses in testicular cells (Zhao et al. 2014). Leydig cells are the majority of somatic cells that locate in the interstitial spaces of the testis. Leydig cells produce androgens essential for normal spermatogenesis and the functions of various organs besides the testis that are targeted by androgens. Leydig cells also play important roles in regulating testicular immune homeostasis.

3. Testicular Immune Privilege

Immune privilege is a term that describes a reduced immune response in certain organs to protect specific vital tissues from an adverse immune response. Typical immunoprivileged organs in mammals include the pregnant uterus, eye, brain, and testis (Stein-Streilein and Caspi 2014). The mammalian testis represents a remarkable immunoprivileged organ considering its tolerance to immunogenic auto- and allo-antigens (Fijak and Meinhardt 2006). The phenotype of testicular immune privilege was observed as early as 1767 when the cock testis was transplanted into the belly of a hen and remained a normal structure for a prolonged period (Setchell 1990). However, the systematic study of the immunoprivileged status in the testis was initiated in the 1970s and 1980s, and it was then found that allografts and xenografts survived for a prolonged time in the testis (Head, Neaves, and Billingham 1983). Another feature of the testicular immune privilege is the tolerance to immunogenic male germ cells. Most male germ cells are produced in the testes after puberty when the central immune tolerance has been established. Therefore, certain germ cell antigens are immunogenic and may induce autoimmune responses if these autoantigens are released into the extra-testis sites. Male germ cells do not induce detrimental immune responses in the genital tracts under physiological conditions due to the special immunological environment of the reproductive system. Intensive studies have been progressively revealing the mechanisms underlying testicular immune privilege (Li, Wang, and Han 2012). The immunoprivileged status in the testes was initially thought to be due to the absence of lymphatic drainage. However, afferent lymphatic vessels have been found in the testes. The sequestration of autoantigens from the immune components by the blood-testes barrier (BTB) was believed to be critical for maintaining testicular immune privilege. The BTB cannot be the only mechanism behind the immunoprivileged status in the whole testis because the interstitial spaces and early stage of germ cells outside the BTB also benefit from immune privilege (Bucy, Chen, and Cooper 1990). An increasing body of evidence shows that a dense network of an immunosuppressive milieu is involved in the regulation of the testicular immune privilege (Figure 2). Leydig cells in the interstitial spaces produce various immunoregulatory factors that play important roles in inhibiting immune responses in the testis and favoring the immunoprivileged status.

4. Role of Leydig Cells in Testicular Immune Privilege

In addition to their endocrine function, Leydig cells regulate the testicular immunoprivileged status through various mechanisms. Leydig cells affect the number and function of the immune cells in the testis. Although the testis is an immunoprivileged organ, various immune cell types reside in the interstitial spaces of the testis under physiological conditions (Figure 1). Macrophages make up the majority of immune cells in the testes. The testicular macrophages predominantly secrete immunosuppressive factors, thereby favoring the testicular immunoprivileged status (Winnall, Muir, and Hedger 2011). The minor other immune cell types, including dendritic cells, lymphocytes, and mast cells, have been found in normal testes, and their functions are not obvious under physiological conditions. However, immune cell numbers can be significantly increased under inflammatory conditions and are involved in the pathogenesis of germ cells (Perez et al. 2013). Leydig cells regulate the number and function of testicular macrophages and lymphocytes in the testis (Duckett et al. 1997, Hedger and Meinhardt 2000). Leydig cells and macrophages are closely associated and interact, thus regulating each other's development. Leydig cells regulate macrophage expansion and maturation depending on the luteinizing hormone (LH) levels. At the endocrine level, Leydig cells also regulate T cell numbers. However, the testicular immune cell numbers cannot be directly regulated by hormones because the immune cells lack hormone receptors. By contrast, Leydig cells could control the immune cell numbers by regulating cell proliferation and migration, but the underlying mechanisms remain to be clarified.

Leydig cells synthesize testosterone under LH regulation. Administration of testosterone suppresses autoimmune diseases (Cutolo 2009). The immunosuppressive properties of testosterone contribute to differences in autoimmune diseases between males and females (Cutolo et al. 2004). Testosterone reduces Toll-like receptor (TLR) 4 expression in macrophages (Rettew, Huet-Hudson, and Marriott 2008). The inhibition of testosterone production reduces regulatory T-cell numbers and increases natural killer cell numbers in human males (Page et al. 2006). These observations indicate that testosterone plays a role in inhibiting autoimmune responses and favoring self-tolerance. Notably, the testosterone level in the testis is 10-fold higher than in normal circulation. Therefore, testosterone should be involved in maintaining a testicular immunoprivileged status. Direct evidence for the association between androgens and testicular immune privilege came from the analysis of mice whose androgen receptors in Sertoli cells were removed. The deletion of the androgen receptor in mouse Sertoli cells compromised the testicular immune privilege (Meng et al. 2011), which has been attributable to the impairment of BTB integrity (Meng et al. 2005). Androgens regulate the expression of various tight junction proteins in Sertoli cells and are essential for the BTB integrity. Therefore, androgens produced by Leydig cells play critical roles in the maintenance of testicular immunoprivileged status by regulating the BTB integrity and inhibiting autoimmune responses.

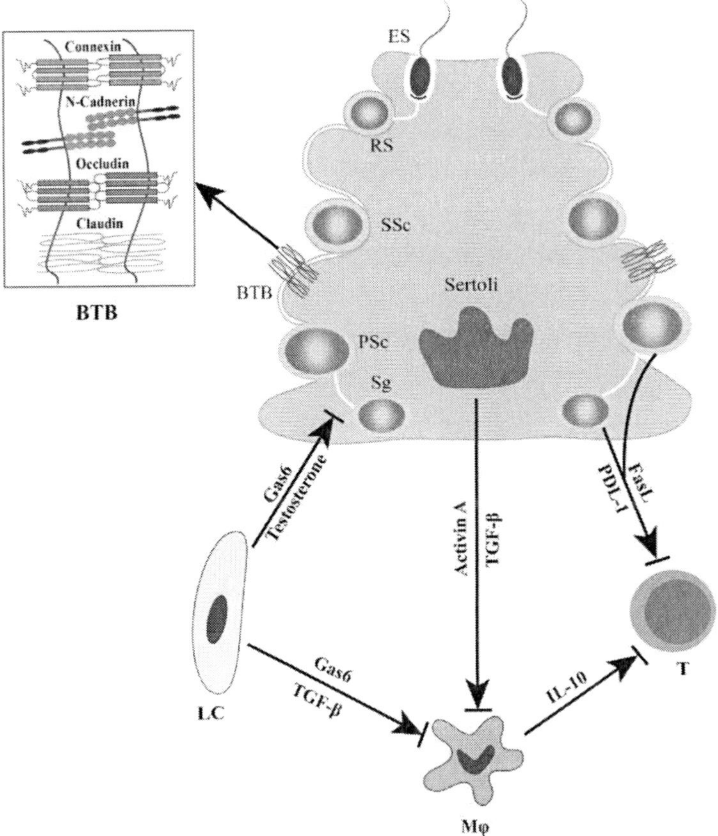

Figure 2. Mechanisms underlying testicular immune privilege. The immunoprivileged status in the testis is maintained by the unique structure blood-testis barrier (BTB) and a network of paracrine immunosuppressive cytokines. The BTB that is created by several types of junctional proteins sequesters most of germ cell antigens from the immune components. Leydig cells (LC) produce Growth arrest-specific gene 6 (Gas6), transforming growth factor-β (TGF-β) and testosterone, which inhibit innate immune responses in macrophages (Mφ) and Sertoli cells. Sertoli cells express activin A and TGF-β that can inhibit immune response of Mφ. Testicular Mφ produce high level of anti-inflammatory cytokine IL-10, thereby inhibiting T lymphocyte activation. Moreover, male germ cells produce soluble FasL and PD-L1, which can inhibit T cell response by inducing cell apoptosis.

In addition to testosterone synthesis, Leydig cells express the immune negative regulatory system. Several immune negative regulatory systems have been found, and these systems play critical roles in preventing autoimmune responses and maintaining immune homeostasis. The Factor associated suicide/Fas ligand (Fas/FasL) system suppresses immune responses by inducing lymphocyte apoptosis (Yamada et al. 2017). FasL is abundantly expressed in the testes (Suda et al. 1993). Although the Fas/FasL system was demonstrated to be critical for the testicular immune privilege by inducing lymphocyte apoptosis via FasL expressed in Sertoli cells (Bellgrau et al. 1995), this conclusion has been challenged by the observation that the neutralizing antibodies against FasL do not compromise the testicular immunoprivileged status (Korbutt et al. 2000). In fact, FasL is predominantly expressed in male germ cells, but not in Sertoli cells (D'alessio et al. 2001). The function of the FasL expressed in male germ cells remains unclear. Another immune checkpoint inhibitory system is the programmed death receptor-1 (PD-1) and its ligand PD-L1. The PD-1/PD-L1 system regulates T cell tolerance via inhibiting T-cell activation. PD-L1 is constitutively expressed in the testis and favors the survival of allografts, indicating that the PD-1/PD-L1 system should be one of the mechanisms underlying testicular immune privilege (Cheng et al. 2009).

Tyro 3, Axl and Mer, collectively termed TAM, belong to a subfamily of receptor tyrosine kinases. One of the most notable functions of TAM receptors is their effect in immune homeostasis (Rothlin et al. 2015). Growth arrest-specific gene 6 (Gas6) is a common functional ligand of TAM receptors. TAM receptors inhibit innate immune responses and promote the phagocytic elimination of apoptotic cells, thereby favoring the resolution of inflammatory conditions and preventing autoimmune diseases (Lemke 2013). TAM receptors and Gas6 are abundantly expressed in the testis and play important roles in regulating the testicular immune privilege (Deng, Chen, and Han 2016). All three TAM receptors are expressed in Sertoli cells. Axl and Mer are also expressed in Leydig cells. Gas6 is uniquely expressed by Leydig cells. However, male germ cells do not express TAM receptors and Gas6. TAM triple knockout ($TAM^{-/-}$) male mice are infertile and progressively lose male germ cells, suggesting that TAM receptors play critical roles in maintaining the testicular function (Lu et al. 1999, Chen et al. 2009). TAM$^{-/-}$ mice develop spontaneous autoimmune orchitis as mice age, suggesting that TAM receptors are essential for the testicular immune homeostasis (Zhang et al. 2013). Therefore, Leydig cells should play a central role in maintaining immune homeostasis by producing Gas6 in the testis. Major testicular cells, including Leydig, Sertoli, and germ cells, are well equipped with PRRs, which initiate innate immune responses and produce various pro-inflammatory factors and chemokines that may facilitate inflammatory conditions. Notably, Gas6 inhibits innate immune responses in both Leydig and Sertoli cells (Sun et al. 2010, Shang et al. 2011). Gas6 significantly inhibits the expression of pro-inflammatory cytokines and chemokines in Sertoli and Leydig cells after stimulation with TLR ligands. In addition to the local inhibitory effect of the TAM/Gas6 system on the inflammatory responses in the testis, this system also inhibits the systemic immune response to male germ cell antigens. While TAM$^{-/-}$ mice develop autoimmune orchitis, Axl and Mer double knockout (Axl$^{-/-}$Mer$^{-/-}$) male mice do not develop spontaneous autoimmune orchitis (Zhang et al. 2013). However, Axl$^{-/-}$Mer$^{-/-}$ male mice are susceptible to the induction of experimental autoimmune orchitis (EAO) (Li et al. 2015). In comparison to wild mice that develop EAO only after triple internal immunization with germ antigens, a single immunization induces severe EAO in Axl$^{-/-}$Mer$^{-/-}$ mice. These observations suggest that TAM receptors regulate the systemic immune tolerance to male germ cell antigens. Therefore, the systemic tolerance to germ cell antigens is also involved in the testicular immunoprivileged status. The mechanisms by which TAM receptors regulate the local immune environment in the testis and the systemic immune response to male germ cell antigens are interesting topics for future investigation, in which the focus should be on the role of Leydig cells in maintaining the testicular immunoprivileged status by producing androgens and Gas6.

5. The Testicular Innate Defense System

Although the testis is a remarkable immunoprivileged organ, it is not sterile. In fact, the testis can be infected by various microbial types, including viruses, bacteria, fungi and parasites, through hematogenous dissemination and ascending male genitourinary tracts. In particular, various types of virus predominantly infect the testis (Dejucq and Jegou 2001). To overcome the immunoprivileged environment and elicit a local response against microbial infections, the testis adopts its own effective innate defense system for counteracting invading pathogens. Although certain types of immune cells reside in the interstitial spaces of the testis, most immune cells display an immunosuppressive phenotype to favor the immunoprivileged status (Bhushan and Meinhardt 2017). By contrast, the tissue-specific cells, including Leydi, Sertoli, and germ cells, initiate their innate defense responses against invading microbes through different mechanisms. Among them, Leydig cells are well equipped with the PRRs that recognize viruses and trigger innate antiviral responses.

5.1. PRR-Initiated Innate Immune Responses

PRRs belong to a superfamily of receptors that can recognize conserved molecular patterns of microbial pathogens, namely, pathogen-associated molecular patterns. PRRs can rapidly initiate innate immune responses against microbial infection (Kumar, Kawai, and Akira 2011). PRRs also recognize endogenous autoantigens that can be released from damaged tissues and necrotic cells, termed damage-associated molecular patterns and trigger noninfectious inflammation (Carta et al. 2009). Several PRR subfamilies have been identified in mammals. Notably, Leydig cells express various PRRs that initiate a different innate immune signaling pathway (Figure 3). TLRs belong to a subfamily of transmembrane proteins, which include 13 members in mammals (O'Neill, Golenbock, and Bowie 2013). TLR signaling pathways have been well characterized. Most TLRs exclusively initiate the myeloid differentiation protein 88 (MyD88)-

dependent pathway except for TLR3 and TLR4. TLR3 activation triggers the Toll/IL-1R domain-containing adaptor inducing IFN-β (TRIF)-dependent pathway, whereas TLR4 initiates both MyD88- and TRIF-dependent pathways. MyD88 signaling results in the activation of NF-κB and mitogen-activated protein kinases (MAPKs), thereby inducing the expression of numerous immunoregulatory cytokines, including pro-inflammatory factors and chemokines. The TRIF-dependent pathway activates NF-κB and interferes regulating factor 3 (IRF3), thus inducing the expression of type 1 interferons (IFN-α/IFN-β) and various proinflammatory cytokines.

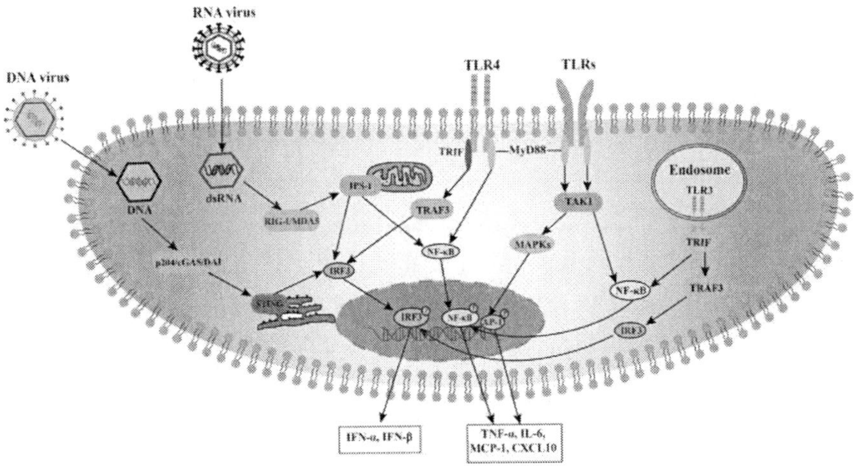

Figure 3. Schematic of PRR signaling in Leydig cells. Most TLRs initiate signaling pathways using adaptor protein MyD88 except for TLR3 and TLR4. TLR3 activation triggers TRIF-dependent pathway, whereas TLR4 initiates both MyD88- and TLIF-dependent pathways. Cytosolic RNA sensors RIG-I and MDA5 recognize viral RNA and trigger innate antiviral response using IPS-1 as an adaptor located on mitochondria. DNA sensor p204 recognizes viral DNA and triggers antiviral response using adaptor protein STING on endoplasmic reticulum. Most PRR signaling pathways activate NF-κB and MAPKs, thus inducing the expression of pro-inflammatory cytokines and chemokines. TRIF, IPS-1 and STING pathways predominantly activate IRF3, which induce IFN-α and IFN-β.

In addition to TLRs, various PRR subfamilies have been identified in the cytoplasm. PRRs recognizing cytosolic double-strained RNA (dsRNA), include two functional members: retinoic acid-inducible gene I (RIG-I) and melanoma differentiation-associated protein 5 (MDA5) (Ori, Murase, and Kawai 2017). RIG-I and MDA5 trigger signaling through IFN-β promoter stimulator-1 (IPS-1). The IPS-1-dependent signaling activates IRF3 and NF-κB, thereby inducing the expression of IFN-α/ IFN-β and pro-inflammatory cytokines. A subfamily of cytosolic DNA sensors have been unraveled (Chen, Sun, and Chen 2016). DNA sensors initiate a signaling pathway using a stimulator of IFN gene (STING) as an adaptor located in the endoplasmic reticulum. STING-dependent DNA sensor signaling pathway predominantly activates IRF3 and induces IFN expression, a universal innate antiviral mechanism in response to viral infection.

5.2. TLRs in the Testis

TLRs were initially thought to be expressed in innate immune cells, including dendritic cells and macrophages. More recent studies found that TLRs and other PRR subfamilies, including cytosolic RNA and DNA sensors, are widely expressed in many types of nonimmune cells. In particular, TLRs and nucleic acid sensors are abundantly expressed in the epithelial cells of the tissues that are frequently infected by microorganisms, such as the intestine, lung, and urogenital tract. Most testicular cells are equipped with PRRs (Figure 4.). The first study on PRRs in the testis investigated TLR expression and function in mouse Sertoli cells and demonstrated that TLR2 and TLR4 signaling pathways were initiated by their ligands (Riccioli et al. 2006). Further studies have shown that murine Sertoli cells express additional functional TLRs, including TLR3, TLR5, and TLR6 (Wu et al. 2008, Starace et al. 2008). After stimulation with the respective ligands, TLRs can be activated in Sertoli cells and trigger testicular innate immune responses. TLR signaling induces the expression of numerous immunoregulatory cytokines, including pro-inflammatory cytokines, chemokines, and type 1 IFNs in Sertoli cells.

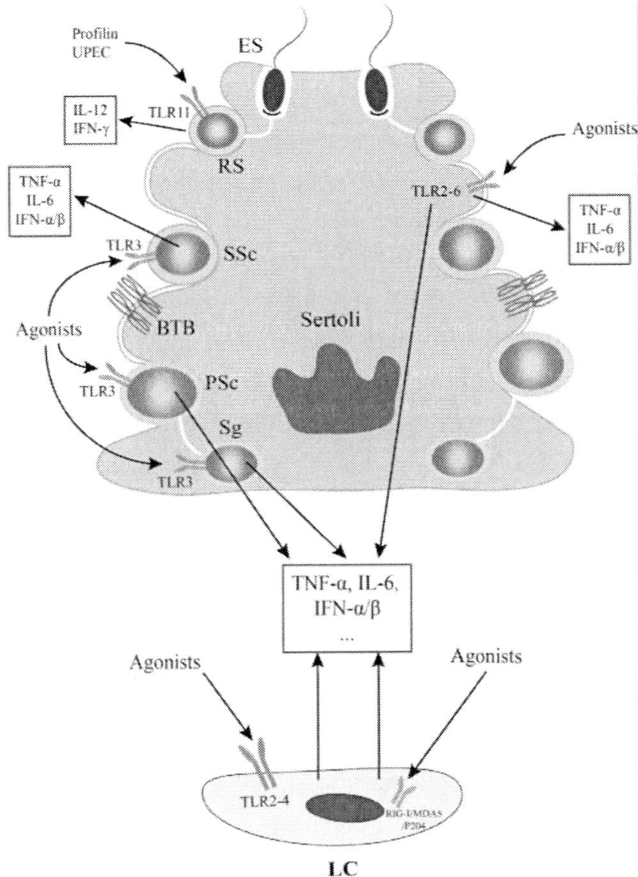

Figure 4. Testicular defense against microbial infection. Major testicular cells abundantly express various PRRs, which can be activated by their respective agonists, thereby initiating innate immune responses to counteract invading microorganisms. Leydig cells (LC) express various PRRs, including Toll-like receptors (TLR 2-4), cytosolic RNA sensors RIG-I and MDA5, and DNA sensor p204. The activation of these PRRs induces the expression of inflammatory cytokine, interferons, and antiviral proteins in Leydig cells. Therefore, Leydig cells play important roles in the testicular defense against microbial infections.

Male germ cells represent the majority of testicular cells during adulthood. These germ cells can be infected by microbial pathogens from the circulating blood and the ascending genital tracts. After stimulation with its ligand, TLR3 initiates innate antiviral responses in mouse spermatogonia and spermatocytes (Wang et al. 2012), two types of male germ cells that reside outside and behind the BTB, respectively. TLR3-initiated antiviral responses should contribute to the germ cell defense against viral infections both outside and behind the BTB. Moreover, a functional TLR11 is expressed in spermatids (Chen et al. 2014). TLR11 recognizes components of *Toxoplasma gondii* and *uropathogenic Escherichia coli* (UPEC), and subsequently induces a cellular response against microbial pathogens. Considering that spermatids reside in the adluminal compartments of the seminiferous epithelium where *T. goddii* and bacteria may infect via the ascending genital tracts, the functional PRRs in spermatids would be particularly important in the defense against invading microbes. However, the functional TLR11 is absent in human beings. The species difference in TLR11 expression may explain the reason why *T. goddii* and UPEC cause pathogenesis in mens' reproductive system, whereas natural *T. goddii* and UPEC infections have not been reported in mice.

Although a large number of macrophages reside in the interstitial spaces of the testis, these macrophages are not well equipped with PRR machineries like their counterparts in the immune system and other organs (Bhushan and Meinhardt 2017). Testicular macrophages express low levels of TLR signaling genes and high levels of inhibitory genes of the TLR signaling pathways (Bhushan et al. 2015). Accordingly, the testicular macrophages belong to subpopulations that predominantly display immunosuppressive properties by producing high levels of anti-inflammatory factors (Gordon 2003). These properties of testicular macrophages contribute to the immunoprivileged status in the testis. In contrast, Leydig cells exhibit a more active innate defense against viral infection than testicular macrophages.

5.3. Innate Defense of Leydig Cells

As a major cell population in the interstitial spaces, Leydig cells play a central role in the testicular defense against invading microbes via hematogenous dissemination. The antiviral responses of testicular cells were unraveled two decades ago (Dejucq, Chousterman, and Jegou 1997). Leydig cells, testicular macrophages, and spermatogonia constitutively express IFNs (Dejucq et al. 1998). In response to viral infection (Le Tortorec et al. 2008), Leydig cells express IFN-inducible antiviral proteins, including IFN-stimulating gene 15 (ISG15), 2'5'-oligodenylate synthetase (OAS1), and MxGTPase 1 (Mx1). ISG15, OAS1, and Mx1 can amplify antiviral signaling, degrade viral RNA and block viral gene transcription, respectively, thereby restricting viral replication within infected cells (Sadler and Williams 2008). Notably, human Leydig cells express relatively low levels of IFNs and antiviral proteins compared to rat Leydig cells, suggesting that the human testis has a weaker antiviral defense than the murine testis (Le Tortorec et al. 2008). This finding may explain the reason why certain viruses induce orchitis in humans and viral orchitis in mice has not been reported. Even when mice are experimentally inoculated with viruses, orchitis is rarely induced. Further understanding of the mechanisms underlying the difference in antiviral responses between human and murine Leydig cells may aid in antiviral therapy and prevention of viral orchitis.

The roles of Leydig cells in the testicular defense against microbial infections have been intensively studied based on the examination of PRRs. Mouse Leydig cells abundantly express TLR3 and TLR4, which can be activated by their respective ligands (Shang et al. 2011). The activation of TLR3 and TLR4 initiates innate immune responses in Leydig cells, thereby leading to the expression of various pro-inflammatory cytokines and IFNs. TLR-initiated innate immune responses suppress testosterone production in Leydig cells, which is presumably an effect of pro-inflammatory cytokines because several of these cytokines, including TNF-α, IL-6, and IL-1, perturb steroidogenesis (Hedger and Meinhardt 2003).

In agreement with their antiviral activities, Leydig cells abundantly express various PRRs that trigger antiviral responses. In addition to the TLR3-initiated innate antiviral response, major cytosolic nucleic acid sensors are abundantly expressed in murine Leydig cells. RIG-I and MDA5 are constitutively expressed in mouse Leydig cells, and their ligand triggers innate antiviral responses via the IPS-1-dependent signaling pathway (Zhu et al. 2013). The IPS-1 signaling pathway results in the activation of NF-κB and IRF3 in Leydig cells, thereby inducing the expression of pro-inflammatory cytokines and type 1 IFNs. Major antiviral proteins, including ISG15, OAS1, and Mx1, were remarkably upregulated in Leydig cells after challenge with RIG-I and MDA5 ligand (Zhu et al. 2013). Notably, the RNA sensor-initiated antiviral responses in Leydig cells inhibit testosterone synthesis, which may perturb testicular functions. Accordingly, RNA viruses, such as human immunodeficiency virus-1(HIV-1), mump virus (MuV) and Zika virus, predominantly induce pathogenesis in the testes (Liu et al. 2018). MuV is a typical virus that frequently induces orchitis, a major etiological factor contributing to male infertility. MuV infects most testicular cells and induces different innate immune responses (Wu et al. 2017). Leydig cells express relatively high levels of IFN-α and IFN-β compared to Sertoli cells in response to MuV infection (Wu et al. 2016). By contrast, Sertoli cells produce high levels of pro-inflammatory cytokines after MuV infection, suggesting that Leydig cells possess more powerful antiviral ability than Sertoli cells, and Sertoli cells should predominantly contribute to the inflammatory condition. Accordingly, MuV replicates at relatively high efficiency in Sertoli cells compared to Leydig cells.

A cytosolic DNA sensor p204 and its signaling adaptor STING are constitutively expressed in mouse Leydig cells (Zhu et al. 2014). The p204 activation induces the expression of IFN-α and IFN-β via STING signaling pathway, which subsequently upregulates antiviral protein expression. In contrast to IPS-1 signaling, the p204-triggered STING pathway in Leydig cells does not alter testosterone synthesis. Accordingly, DNA viruses rarely cause orchitis and barely affect male fertility. Therefore, the manipulation of STING signaling could be an ideal strategy for promoting the testicular defense against viral infection and protecting the testicular function. The differences in testicular pathogenesis in response to RNA and DNA viral infections are worthy of in-depth investigation.

CONCLUSION

The mammalian testis adopts an immunoprivileged environment to protect male germ cells from adverse immune responses. To overcome the immune privilege, the testis forms an effective local innate defense system against microbial infection. The interplay among tissue structure, testicular cells, and cytokine milieu regulates the testicular immune environment and the local defense system. As a major testicular cell population, Leydig cells play important roles in regulating the local immune status and innate defense. Leydig cells are the only testicular cells that produce testosterone and Gas6. Both testosterone and Gas6 suppress immune responses and are essential for the maintenance of the immunoprivileged environment. Leydig cells are also well equipped with PRRs and produce high levels of IFNs in response to viral infection. A broad range of viruses infect the testis. The function of Leydig cells can be affected during an infection as these may inhibit testosterone synthesis, thereby perturbing testicular functions and immune homeostasis. Therefore, the roles of Leydig cells in the pathophysiology of the testes in response to viral infection require an in-depth investigation. In this regard, we should focus on the different responses of Leydig cells to RNA and DNA viral infections because RNA viruses more frequently impair testicular functions than DNA viruses.

ACKNOWLEDGMENTS

This work was supported by grands from CAMS Initiative for Innovative Medicine (Nos. 2017-I2M-B&R-06 and 2017-I2M-3-007), National Key Research & Developmental Program of China (Nos. 2016YFA0101001 and 2018YFC1003902).

REFERENCES

Bellgrau, D., D. Gold, H. Selawry, J. Moore, A. Franzusoff, and R. C. Duke. (1995). A role for CD95 ligand in preventing graft rejection. *Nature*, 377 (6550):630-632.

Bhushan, S., and A. Meinhardt. (2017). The macrophages in testis function. *J Reprod Immunol*, 119:107-112.

Bhushan, S., S. Tchatalbachev, Y. Lu, S. Frohlich, M. Fijak, V. Vijayan, T. Chakraborty, and A. Meinhardt. (2015). Differential activation of inflammatory pathways in testicular macrophages provides a rationale for their subdued inflammatory capacity. *J Immunol*, 194 (11):5455-5464.

Bucy, R. P., C. H. Chen, and M. D. Cooper. (1990). Ontogeny of T cell receptors in the chicken thymus. *J Immunol*, 144 (4):1161-1168.

Carta, S., P. Castellani, L. Delfino, S. Tassi, R. Vene, and A. Rubartelli. (2009). DAMPs and inflammatory processes: the role of redox in the different outcomes. *J Leukoc Biol*, 86 (3):549-555.

Chen, Q., L. Sun, and Z. J. Chen. (2016). Regulation and function of the cGAS-STING pathway of cytosolic DNA sensing. *Nat Immunol*, 17 (10):1142-1149.

Chen, Q., W. Zhu, Z. Liu, K. Yan, S. Zhao, and D. Han. (2014). Toll-like receptor 11-initiated innate immune response in male mouse germ cells. *Biol Reprod*, 90 (2):38.

Chen, Y., H. Wang, N. Qi, H. Wu, W. Xiong, J. Ma, Q. Lu, and D. Han. (2009). Functions of TAM RTKs in regulating spermatogenesis and male fertility in mice. *Reproduction*, 138 (4):655-666.

Cheng, X., H. Dai, N. Wan, Y. Moore, R. Vankayalapati, and Z. Dai. (2009). Interaction of programmed death-1 and programmed death-1 ligand-1 contributes to testicular immune privilege. *Transplantation*, 87 (12):1778-1786.

Cutolo, M. (2009). Androgens in rheumatoid arthritis: when are they effectors? *Arthritis Res Ther*, 11 (5):126.

Cutolo, M., A. Sulli, S. Capellino, B. Villaggio, P. Montagna, B. Seriolo, and R. H. Straub. (2004). Sex hormones influence on the immune system: basic and clinical aspects in autoimmunity. *Lupus*, 13 (9):635-638.

D'Alessio, A., A. Riccioli, P. Lauretti, F. Padula, B. Muciaccia, P. De Cesaris, A. Filippini, S. Nagata, and E. Ziparo. (2001). Testicular FasL is expressed by sperm cells. *Proc Natl Acad Sci U S A*, 98 (6):3316-3321.

Dejucq, N., S. Chousterman, and B. Jegou. (1997). The testicular antiviral defense system: localization, expression, and regulation of 2'5' oligoadenylate synthetase, double-stranded RNA-activated protein kinase, and Mx proteins in the rat seminiferous tubule. *J Cell Biol*, 139 (4):865-873.

Dejucq, N., and B. Jegou. (2001). Viruses in the mammalian male genital tract and their effects on the reproductive system. *Microbiol Mol Biol Rev*, 65 (2):208-231.

Dejucq, N., M. O. Lienard, E. Guillaume, I. Dorval, and B. Jegou. (1998). Expression of interferons-alpha and -gamma in testicular interstitial tissue and spermatogonia of the rat. *Endocrinology*, 139 (7):3081-3087.

Deng, T., Q. Chen, and D. Han. (2016). The roles of TAM receptor tyrosine kinases in the mammalian testis and immunoprivileged sites. *Front Biosci (Landmark Ed)*, 21:316-327.

Duckett, R. J., N. G. Wreford, S. J. Meachem, R. I. McLachlan, and M. P. Hedger. (1997). Effect of chorionic gonadotropin and flutamide on Leydig cell and macrophage populations in the testosterone-estradiol-implanted adult rat. *J Androl*, 18 (6):656-662.

Fijak, M., and A. Meinhardt. (2006). The testis in immune privilege. *Immunol Rev*, 213:66-81.

Gordon, S. (2003). Alternative activation of macrophages. *Nat Rev Immunol*, 3 (1):23-35.

Head, J. R., W. B. Neaves, and R. E. Billingham. (1983). Immune privilege in the testis. I. Basic parameters of allograft survival. *Transplantation*, 36 (4):423-431.

Hedger, M. P., and A. Meinhardt. (2000). Local regulation of T cell numbers and lymphocyte-inhibiting activity in the interstitial tissue of the adult rat testis. *J Reprod Immunol*, 48 (2):69-80.

Hedger, M. P., and A. Meinhardt. (2003). Cytokines and the immune-testicular axis. *J Reprod Immunol*, 58 (1):1-26.

Korbutt, G. S., W. L. Suarez-Pinzon, R. F. Power, R. V. Rajotte, and A. Rabinovitch. (2000). Testicular Sertoli cells exert both protective and destructive effects on syngeneic islet grafts in non-obese diabetic mice. *Diabetologia*, 43 (4):474-480.

Kumar, H., T. Kawai, and S. Akira. (2011). Pathogen recognition by the innate immune system. *Int Rev Immunol*, 30 (1):16-34.

Le Tortorec, A., H. Denis, A. P. Satie, J. J. Patard, A. Ruffault, B. Jegou, and N. Dejucq-Rainsford. (2008). Antiviral responses of human Leydig cells to mumps virus infection or poly I:C stimulation. *Hum Reprod*, 23 (9):2095-2103.

Lemke, G. (2013). Biology of the TAM receptors. *Cold Spring Harb Perspect Biol*, 5 (11):a009076.

Li, N., Z. Liu, Y. Zhang, Q. Chen, P. Liu, C. Y. Cheng, W. M. Lee, Y. Chen, and D. Han. (2015). Mice lacking Axl and Mer tyrosine kinase receptors are susceptible to experimental autoimmune orchitis induction. *Immunol Cell Biol*, 93 (3):311-320.

Li, N., T. Wang, and D. Han. (2012). Structural, cellular and molecular aspects of immune privilege in the testis. *Front Immunol*, 3:152.

Liu, W., R. Han, H. Wu, and D. Han. (2018). Viral threat to male fertility. *Andrologia*, 50 (11):e13140.

Lu, Q., M. Gore, Q. Zhang, T. Camenisch, S. Boast, F. Casagranda, C. Lai, M. K. Skinner, R. Klein, G. K. Matsushima, H. S. Earp, S. P. Goff, and G. Lemke. (1999). Tyro-3 family receptors are essential regulators of mammalian spermatogenesis. *Nature*, 398 (6729):723-728.

Meng, J., A. R. Greenlee, C. J. Taub, and R. E. Braun. (2011). Sertoli cell-specific deletion of the androgen receptor compromises testicular immune privilege in mice. *Biol Reprod*, 85 (2):254-260.

Meng, J., R. W. Holdcraft, J. E. Shima, M. D. Griswold, and R. E. Braun. (2005). Androgens regulate the permeability of the blood-testis barrier. *Proc Natl Acad Sci U S A*, 102 (46):16696-16700.

O'Neill, L. A., D. Golenbock, and A. G. Bowie. (2013). The history of Toll-like receptors - redefining innate immunity. *Nat Rev Immunol*, 13 (6):453-460.

Ori, D., M. Murase, and T. Kawai. (2017). Cytosolic nucleic acid sensors and innate immune regulation. *Int Rev Immunol*, 36 (2):74-88.

Page, S. T., S. R. Plymate, W. J. Bremner, A. M. Matsumoto, D. L. Hess, D. W. Lin, J. K. Amory, P. S. Nelson, and J. D. Wu. (2006). Effect of medical castration on CD4+ CD25+ T cells, CD8+ T cell IFN-gamma expression, and NK cells: a physiological role for testosterone and/or its metabolites. *Am J Physiol Endocrinol Metab*, 290 (5):E856-863.

Perez, C. V., M. S. Theas, P. V. Jacobo, S. Jarazo-Dietrich, V. A. Guazzone, and L. Lustig. (2013). Dual role of immune cells in the testis: Protective or pathogenic for germ cells? *Spermatogenesis*, 3 (1):e23870.

Rettew, J. A., Y. M. Huet-Hudson, and I. Marriott. (2008). Testosterone reduces macrophage expression in the mouse of toll-like receptor 4, a trigger for inflammation and innate immunity. *Biol Reprod*, 78 (3):432-437.

Riccioli, A., D. Starace, R. Galli, A. Fuso, S. Scarpa, F. Palombi, P. De Cesaris, E. Ziparo, and A. Filippini. (2006). Sertoli cells initiate testicular innate immune responses through TLR activation. *J Immunol*, 177 (10):7122-7130.

Rothlin, C. V., E. A. Carrera-Silva, L. Bosurgi, and S. Ghosh. (2015). TAM receptor signaling in immune homeostasis. *Annu Rev Immunol*, 33:355-391.

Sadler, A. J., and B. R. Williams. (2008). Interferon-inducible antiviral effectors. *Nat Rev Immunol*, 8 (7):559-568.

Setchell, B. P. (1990). The testis and tissue transplantation: historical aspects. *J Reprod Immunol*, 18 (1):1-8.

Shang, T., X. Zhang, T. Wang, B. Sun, T. Deng, and D. Han. (2011). Toll-like receptor-initiated testicular innate immune responses in mouse Leydig cells. *Endocrinology*, 152 (7):2827-2836.

Starace, D., R. Galli, A. Paone, P. De Cesaris, A. Filippini, E. Ziparo, and A. Riccioli. (2008). Toll-like receptor 3 activation induces antiviral immune responses in mouse sertoli cells. *Biol Reprod*, 79 (4):766-775.

Stein-Streilein, J., and R. R. Caspi. (2014). Immune privilege and the philosophy of immunology. *Front Immunol*, 5:110.

Suda, T., T. Takahashi, P. Golstein, and S. Nagata. (1993). Molecular cloning and expression of the Fas ligand, a novel member of the tumor necrosis factor family. *Cell*, 75 (6):1169-1178.

Sun, B., N. Qi, T. Shang, H. Wu, T. Deng, and D. Han. (2010). Sertoli cell-initiated testicular innate immune response through toll-like receptor-3 activation is negatively regulated by Tyro3, Axl, and mer receptors. *Endocrinology*, 151 (6):2886-2897.

Wang, T., X. Zhang, Q. Chen, T. Deng, Y. Zhang, N. Li, T. Shang, Y. Chen, and D. Han. (2012). Toll-like receptor 3-initiated antiviral responses in mouse male germ cells in vitro. *Biol Reprod*, 86 (4):106.

Winnall, W. R., J. A. Muir, and M. P. Hedger. (2011). Rat resident testicular macrophages have an alternatively activated phenotype and constitutively produce interleukin-10 in vitro. *J Leukoc Biol*, 90 (1):133-143.

Wu, H., L. Shi, Q. Wang, L. Cheng, X. Zhao, Q. Chen, Q. Jiang, M. Feng, Q. Li, and D. Han. (2016). Mumps virus-induced innate immune responses in mouse Sertoli and Leydig cells. *Sci Rep*, 6:19507.

Wu, H., H. Wang, W. Xiong, S. Chen, H. Tang, and D. Han. (2008). Expression patterns and functions of toll-like receptors in mouse sertoli cells. *Endocrinology*, 149 (9):4402-4412.

Wu, H., X. Zhao, F. Wang, Q. Jiang, L. Shi, M. Gong, W. Liu, B. Gao, C. Song, Q. Li, Y. Chen, and D. Han. (2017). Mouse Testicular Cell Type-Specific Antiviral Response against Mumps Virus Replication. *Front Immunol*, 8:117.

Yamada, A., R. Arakaki, M. Saito, Y. Kudo, and N. Ishimaru. (2017). Dual Role of Fas/FasL-Mediated Signal in Peripheral Immune Tolerance. *Front Immunol*, 8:403.

Zhang, Y., N. Li, Q. Chen, K. Yan, Z. Liu, X. Zhang, P. Liu, Y. Chen, and D. Han. (2013). Breakdown of immune homeostasis in the testis of mice lacking Tyro3, Axl and Mer receptor tyrosine kinases. *Immunol Cell Biol*, 91 (6):416-426.

Zhao, S., W. Zhu, S. Xue, and D. Han. (2014). Testicular defense systems: immune privilege and innate immunity. *Cell Mol Immunol*, 11 (5):428-437.

Zhu, W., Q. Chen, K. Yan, Z. Liu, N. Li, X. Zhang, L. Yu, Y. Chen, and D. Han. (2013). RIG-I-like receptors mediate innate antiviral response in mouse testis. *Mol Endocrinol*, 27 (9):1455-1467.

Zhu, W., P. Liu, L. Yu, Q. Chen, Z. Liu, K. Yan, W. M. Lee, C. Y. Cheng, and D. Han. (2014). p204-initiated innate antiviral response in mouse Leydig cells. *Biol Reprod*, 91 (1):8.

In: Leydig Cells
Editor: Bruno Solomon

ISBN: 978-1-53617-282-9

Chapter 2

LEYDIG CELL LINES: *IN VITRO* MODELS FOR TESTOSTERONE BIOSYNTHESIS

Nebahat Yildizbayrak [*] ***and Melike Erkan***
Department of Biology, Istanbul University, Istanbul, Turkey

ABSTRACT

In recent years, the incidence of male infertility is considered as a global health problem. Testicular damage and functional changes in the reproductive system is great concern for male reproduction. Testes contain several cells types such as Sertoli, Leydig and spermatogenic cells. Leydig cells which of them located in interstitial compartment of testes play an essential role in maintaining spermatogenesis and biosynthesis of testosterone. In the male reproductive system, it is necessary to examine the steroidogenic pathway, through the expression levels of the genes and proteins involved, secondary messengers and many other factors responsible for cell function in order to reveal the conditions affecting androgen production. Today, cell culture studies are very important for animal welfare and provide scientists with the opportunity to investigate the molecular mechanism of diseases, to elucidate cell toxicity and to model models for various clinical problems.

[*] Corresponding Author Email: nebahat.yb@gmail.com.

This chapter focuses on the comparison of commercially available immortalized Leydig cell lines derived from rat or mouse and the results of studies with these cells.

Keywords: Leydig cell, cell culture, steroidogenesis, testosterone

INTRODUCTION

Reproduction is a physiological process that provides genetic continuity in all living organisms. The male reproductive system consists of the testes, epididymis, seminal vesicle prostate gland and the penis. While the testes are responsible for sperm production, the epididymis and genital tracts allow the maturation and storage of the sperm. The seminal vesicle produces and secretes the semen fluid. In addition, the prostate gland and the penis allows the sperm to be transferred to the female tracts [1].

Testis, which is a part of male reproductive system and endocrine system, has two main functions. The first of these; male germ cell production for genetic continuity of species and the second of them is the production and secretion of steroid hormones that are necessary for male sexual differentiation and development of secondary sex characters [2]. Testis is divided anatomically into two parts as seminiferous tubules and interstitial space. Seminiferous tubules form the non-vascular testicular area, including Sertoli cells and germ cells which are present at various stages of spermatogenesis [3]. Sertoli cells which play an important role in the regulation of spermatogenesis in seminiferous epithelium; serves as a vital support for the control of spermatogonium, the development and the survival of sperm [4]. In addition, neighbouring Sertoli cells establish close connections with each other to form the blood-testicular barrier. The area between the testicular seminiferous tubules is called interstitial area. Interstitial area is composed of Leydig cells, macrophages, fibroblasts, myeloid cells and connective tissue which is rich in blood and lymph vessels [5].

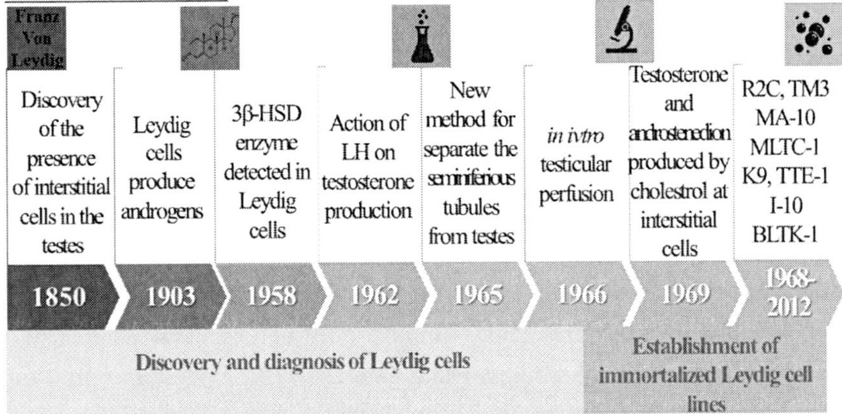

Figure 1. Milestones of Leydig cell research.

The historical starting point of Leydig cell research is regarded as Franz Leydig's description of cells in mammalian testes [6]. In the early 1900s, Pol Bouin and Paul Ancel first time reported that Leydig cells had an endocrine function [7-10]. After understanding the effect of luteinizing hormone on androgen endocrinology, it was concluded that Leydig cells were responsible for testicular androgenic hormone secretion and histological staining of 3β-hsd was recorded as the first histochemical evidence [11]. In the following few years, the first biochemical finding that the interstitial area of the testis was the main source of androgen production was proved by Christensen and Mason in 1965 [12]. With the testicular perfusion methodology developed by Larry Ewing in 1966, Leydig cell function studies gained speed and new information was reached about the steroidogenic pathway [13, 14]. The study by Hall et al. showed that testicular androgens are produced in Leydig cells by metabolizing of cholesterol [15].

Leydig cells, are the main source of testosterone, which is the androgen necessary for the control of spermatogenesis, male reproductive tract and the development of secondary sex characteristics. In male, androgen production is controlled by different mechanisms during foetal and adult periods, and these pathways associated with two different Leydig cell populations called foetal Leydig cells and adult Leydig cells [16]. These two populations show differences in histological structure, life span, androgen synthesis capacity and responsiveness to gonadotropins. Foetal Leydig cells, the main source of androgen production during the prenatal period, are involved in the masculinization of internal and external male genital structures and regulation of neuroendocrine functions [17]. Although the source of foetal Leydig cells is still unknown, they might be originated from adrenal gonadal primordium, neural crest mesonephros and coelomic epithelium. Foetal Leydig cells are morphologically round-oval, have smooth endoplasmic reticulum, mitochondria and abundant lipid droplets also they are in tendency to make group and are surrounded by a membrane consisting of collagen and laminin [18]. Foetal Leydig cells start testosterone production in rats at gestational day 15.5 and continue actively until birth. However, in mice, androstenedione production by foetal Leydig cells occurs from gestational day 12.5 and testosterone conversion is controlled by foetal Sertoli cells [19].

Adult type Leydig cells, which arise from differentiation of mesenchymal cells in testicular development, begin to be seen in postnatal day 10 and increase their numbers gradually after day 15 [20]. They develop as broad, round or polygonal and have an eosinophilic cytoplasm with a centrally located nucleus [5]. Adult Leydig cells play a role in initiation of spermatogenesis, and regulation of endocrine functions in puberty and the production of testosterone for pubertal development of the external genitalia [17]. Unlike foetal Leydig cells, they can aggregate without being surrounded by a membrane and have a higher potential for producing androgens [21]. Both Leydig cell populations contain lipase, oxidative enzymes, esterase and steroid dehydrogenase enzymes in their cytoplasm and they can express the luteinizing hormone receptor [22]. The most critical point in which two Leydig cell populations differ functionally is that luteinizing hormone is responsible for the development and regulation of steroidogenic function of adult Leydig cells, whereas foetal Leydig cells do not require luteinizing hormone [20].

Stimulation of steroidogenesis and secretion of testicular androgens in mammals depends on the number and activity of Leydig cells. Testosterone, the main androgen, is the hormone essential for sexual development and descent of the testes during foetal period, sperm production in seminiferous tubules, maintaining of accessory sex organs and sexual behaviour during puberty [23-25]. Testosterone biosynthesis is mainly controlled by the luteinizing hormone (LH), a gonadotropin released from the pituitary gland. LH, which binds to the specialized receptor (G protein coupled LH human chorionic gonadotropin (hCG) receptor) on the surface of Leydig cells, triggers the intracellular signal cascade for steroidogenesis. The first response to LH is the production of 3,' 5'-cylic adenosine monophosphate (cAMP), the intracellular secondary messenger, which has two major roles in the steroidogenesis process in Leydig cells. The first is acute regulation of testosterone biosynthesis by transporting cholesterol into steroidogenic pathway. cAMP-dependent protein kinase A activates the transport of cholesterol from intercellular lipoprotein sources, from de nova synthesis of cholesterol or from intracellular cholesterol pools [23]. The transport of cholesterol into the mitochondrial inner membrane is a cAMP-dependent process but takes place under the control of steroidogenic acute regulatory protein (StAR), this translocation is defined as the rate-limiting step of steroidogenesis. The second and chronic effect of cAMP on Leydig cells is regulation of the gene expression of cytochrome P450 proteins in mitochondria and hydroxysteroid dehydrogenases in smooth endoplasmic reticulum [26].

Testosterone biosynthesis begins with the conversion of cholesterol into pregnenolone through the cholesterol side chain cleavage (Cyp11a1, P450scc) enzyme found in the mitochondrial inner membrane. After pregnenolone passes to the smooth endoplasmic reticulum where progesterone is formed by the action of 3β-hydroxy steroid dehydrogenase (3β-HSD). Progesterone is then converted to 17α-hydroxy progesterone and then to androstenedione via the 17α-hydroxylase/C17-20 lyase enzyme (Cyp17a1). In the last step of steroidogenesis, androstenedione forms testosterone with the action of 17β-hydroxysteroid-dehydrogenase (17β-HSD) enzyme [25, 27].

Immortalize Cell Lines

Increasing industrialization in the world causes people to be exposed to many synthetic chemicals throughout their life cycles. The incidence of reproductive disorders has increased considerably in recent years due to increased environmental pollutants, pharmaceuticals and cosmetic products used and changing eating habits. Treatment of various metabolism and hormone disorders associated with decreased serum testosterone levels and in order to find pharmacological solutions, the examination of testosterone production mechanism is critical for male reproductive system. In addition, it is important to examine steroidogenesis *in vitro* in order to detect potential reproductive toxicants. Nowadays the use of cell cultures in combination with developing *in vitro* diagnosis methods is very common in the early stages of drug research and in the investigation of molecular pathways and developmental cascades [28]. In endocrinology studies, cell cultures are indispensable for investigating hormone receptor complexes and for determining gene-protein expression by reducing dependence on animal experiments. The use of immortalized cell lines provides several advantages over primary cell cultures due to their unlimited lifespan properties, easier maintenance and continuity in long-term experiments. However, immortalized cell lines need to be well characterized and it is essential to work under aseptic conditions [29]. The human adrenal H295R cell line, developed for the study of adrenal steroidogenesis, was validated by the Organization for Economic Cooperation and Development (OECD) to reveal the effects of endocrine disrupting chemicals on steroid hormone production [30]. However, since it cannot express 17β-hsd type 3, H295R cell-based system is insufficient to evaluate testosterone production. Although there are many commercial murine Leydig cell lines for the investigation of testicular steroidogenesis, a human Leydig cell line has not yet been developed [31]. The present murine Leydig cell lines show differences in their immortalization patterns, the properties of the tissue from which they are originated, and the steroidogenic enzymes they can express [20]. Each of these features cause some advantages or limits for the immortalized cell lines. When all these are taken into consideration, it is critical to make cell line selection by analysing the endpoints of the study and target molecules correctly in the testosterone production process.

R2C (ATCC, CCL-97) Cell Line

R2C cells were established in 1968, the first immortalized Leydig cell line of stable, clonal and monolayer culture, immortalized from Leydig tumour cells of Rattus norvegicus rats [32]. In addition to 3β-HSD1 and Cyp11a1 expressions, they have been reported to produce testosterone by expressing Cyp17a1. StAR expression and progesterone production in R2C cells are high furthermore cells have a characteristic that makes them different from Leydig cell lines developed in the following years. These cells, which can produce progesterone without hormonal stimulation, have been shown to be capable of steroid secretion independent of cAMP production [33]. After proving the clonal cell lines could be a useful model in studies investigating hormonal biosynthesis and target cell responses, Leydig cell lines from different species and tissues (normal or malignant) continued to be developed.

MA-10 (ATCC, CRL-3050) Cell Line

One of the most studied inbred Mus musculus models, MA-10 Leydig cells originating from C57BL/6 line are tumour Leydig cells in epithelial morphology. MA-10 cells, having a homogeneous cell population, are widely used as they are very similar to Leydig cells of normal character [34, 35]. LHR receptors are present on cell surfaces and they respond to LH/hCG stimulation by producing progesterone. In addition, expression of Cyp17a1 was found to be low in the MA-10 cells in contrast to high activity of 3β-HSD1. Due to their enzymatic profiles of MA-10 cells, their steroidogenesis capacity seems to be partial, however it is a suitable cell line for studying cell function and gonadotropin actions. MA-10 cells were used to investigate the effect of high StAR expression and high cholesterol mobilization on steroid production of R2C cells [33]. To investigate the performance of steroid hormone production, two different groups of MA-10 cells transfected with StAR cDNA or cAMP stimulated and results showed that trophic hormone stimulation is more efficient in productivity of steroidogenesis. Thus, it was concluded that intense StAR expression was not enough to increase steroidogenic capacity [36, 37].

MLTC-1 (ATCC, CRL-2065) Cell Line

MLTC-1, another cell line obtained from C57BL/6 strain, retains the hormonal response characteristics of the tumour tissue from which it originated [38]. Membrane adenyl cyclase activities in MLTC-1 Leydig tumour cell line are regulated by LH and gonadotropin treatment. Since they originate from the same origin MA-10 and MLTC-1 cell lines show very similar characteristics and progesterone production of both cell lines is higher than testosterone production due to the characteristic of the tumour tissue from which they originate [38].

The transformed cell lines do not support the study of complex metabolic cell pathways since they are generally limited to the expression of one specialized protein. The R2C, MA-10 and MLTC-1 transfected tumour Leydig cell lines are therefore unable to reach the final step of the steroidogenic pathway. To overcome this problem, a Leydig cell line developed by the hybridization method of fresh cells isolated from tissue of the same species as the transformed cell line was established [39]. This hybrid cell line, termed K9 by Finaz et al., was obtained from freshly isolated mouse Leydig cells and MA-10 tumour Leydig cells. A disadvantage of this cell line, which can express all the steroidogenic enzymes required for testosterone production, is that since K9 cells are hybridized and lack stability, prolonged passage process resulting in MA-10 cell phenotype [39, 40]. In order to perform efficient and sustainable experiments, R2C cells require regular subcloning.

I10 (ATTC, CCL-83) AND LC-540 (ATTC, CCL-43) CELL LINES

The mechanisms of luteinizing hormone responsiveness and cAMP production of I10 (Mus musculus) and LC-540 (Rattus norvegicus) clonal lines that originated from Leydig tumour cells are not fully understood. Furthermore, although LC-540 cells are known to produce testosterone, it is not clear whether steroid production is induced by gonadotropins. Since these two cell lines are not as well characterized as other clonal cell lines, they are not preferred in the studies [41, 42].

TM3 (ATTC, CRL-1714) CELL LINE

TM3 Leydig cell line was developed by spontaneous immortalization from primary testicular murine cells, unlike clonal immortal cell lines of Leydig cell tumour origin. TM3 cells, the non-tumour Leydig cell line, were isolated from 11-13 days old Mus musculus mice and responded to luteinizing hormone with an increase in cAMP production [43]. Furthermore, TM3 cells are metabolizing cholesterol after treatment with luteinizing hormone. Since the expression of StAR, Cyp11a1 and 3β-HSD1 is active, the final products in steroidogenesis are mainly progesterone. TM3 cells can LHR expression which is specific to Leydig cells as well as secretion of testosterone by calcitonin receptors stimulating calcium influx and cAMP production [43].

TTE-1 CELL LINE

Immortal cell lines were also developed from primary Leydig cell cultures by *in vitro* and *in vivo* Tag transfection method [20]. TTE1 cell line immortalized using temperature-sensitive simian virus 40 (SV40) large T-antigen from transgenic mice is the first cell line established from the transgenic mice. Their ability to express 17β-HSD1 and 17β-HSD3 differs them from the previously established clonal cell lines [44]. TTE1 cells showed temperature sensitive (33°C) growth characteristics and a differentiated phenotype at a nonpermissive temperature of 39°C. In microarray analysis with TTE1 cells cultured at permissive and not permissive temperatures, it was concluded that many genes associated with Leydig cell differentiation and function were expressed [45].

BLT-1 Cell Line

Another immortal Leydig cell line BLT-1 generated from testicular tumour induced with SV40T- antigen under the control of the inhibin-α promotor at transgenic mice [46]. They have high responsiveness to LH/hCG and remain stable even long-term passages, while they can produce high amounts of progesterone and only low amounts of testosterone, as the previously established MA-10 and MLTC-1. Since the number of LH receptors found in BLT-1 cells (36000/cell) is very close to the primary cultures of Leydig cells (42000/cell), cAMP production and steroidogenesis of the BLT-1 cells are performed at a very high performance [47]. In 2002, a new steroidogenic model was developed from BLT-1 cells for evaluate the environmental toxicants on steroidogenesis. BLTK1 murine Leydig cells, BLT-1 derived cell clone, can express all essential steroidogenic enzymes including StAR, Cyp11a1, Cyp17a1, HSD3B, HSD17B and SRD5A1. BLTK1 cells as Leydig cell model stands out with their LH receptors, being very stable, easily transfected and growth characteristics [48].

Recent Literature with Immortalize Cell Lines

Since the decrease in testosterone levels can not only cause male reproductive system disorders but also can trigger metabolic diseases such as cardiovascular diseases and obesity, it is important to investigate the effects of daily exposure toxins on testosterone production [49]. Almost every day a new one is added to the studies which of them used Leydig cell lines as a steroidogenic model in the early stages of endocrine studies and in the investigation of the effects of endocrine disrupting environmental toxins and endogenous contaminants on male reproductive system. Articles published over the last few years have been outlined with a review of Odermatt and colleagues [31]. Although Leydig cell lines in use have advantages and disadvantages compared to each other, this chapter mentioned the studies with most commonly used Leydig cell lines include MA-10, MLTC-1 and TM3.

Non-tumoral TM3 cells have been preferred for revealing the effects of bisphenol A on testosterone production because TM3 cells have both androgen and oestrogen receptors [50]. Bisphenol A is known to have an antagonist interaction with androgen and oestrogen receptors. After bisphenol A treatment in concentrations determined from the serum of the individuals as a result of occupational exposure, testosterone level was determined by using chemiluminescence assay in the experimental groups. The amount of testosterone measured in the collected culture media was suppressed by 20% to 40%, thus the authors concluded that bisphenol A impaired steroidogenic function in TM3 cells [50]. In another study TM3 cell line and primary Leydig cells isolated from adult rats were used in response to *in vitro* and *in vivo* systems for investigating the effects of benzo (a) pyrene on the steroidogenic pathway and the antioxidant properties of resveratrol [51]. StAR, Cyp11a1, 3β-hsd and 17β-hsd expressions, which have critical roles in testosterone production, were analysed by Western blot and real time-PCR methods, also testosterone levels were determined by ELISA. Bisphenol A has been reported to adversely affect Leydig cell steroidogenesis, while resveratrol has been shown to have a protective effect. In this study, the results of *in vivo* and *in vitro* experiments showed consistency with each other, but it was emphasized that the results should be supported with clinical trials in order to determine the antioxidant dose of resveratrol [51]. In another study with TM3 cells, the effects of fluoride, an essential trace element, was investigated at concentrations below the limits permitted by World Health Organization. The results of the study showed that fluoride reduced cAMP levels and disrupted key points in steroidogenesis. Authors also supported the decrease in the amount of testosterone with the findings that steroidogenic genes (StAR, Cyp11a1, 3β-hsd and 17β-HSD) and regulatory transcription factors (steroidogenic factor-1, GATA binding protein-4 and nerve growth factor IB) are negatively affected after exposure to fluoride [52]. In addition to endocrine disrupting chemicals, Leydig cell lines are used to determine the effects of endogenous factors on steroidogenesis. The effects of brain-derived neurotrophic factor on Leydig cell function were investigated in TM3 cell line [53]. As is known, testosterone production in Leydig cells is regulated by paracrine and autocrine factors such as luteinizing hormone and follicle stimulating hormone. In this study, brain-derived neurotrophic factor has been shown to increase testosterone production by increasing the expression of basic steroidogenesis regulatory molecules (StAR, Cyp11a1, 3β-hsd) [53].

MA-10 cells were shown to be complementary to H295R cell line in testicular steroidogenesis examination. MA-10 Leydig cells used as a model to investigate the combined effects of mycotoxins and pesticides on steroidogenesis were analysed primarily in basal and stimulated conditions for progesterone and testosterone production and positive results were obtained for both steroids [54]. Incubation of MA-10 cells with combination of 1,1,1-trichloro-2,2-bis(p-chlorophenyl) ethane, zearalenol, and/or metabolites of zearalenol resulted in decreased testosterone production. In this study, the effects of mycotoxins and pesticides in various combinations on steroid production have been shown by utilizing MA-10 cells to express both oestrogen and progesterone receptors. Decrease in testosterone and increase in progesterone in different combinations have been associated with inhibition of Cyp17a1 enzyme [54]. MA-10 cells were used to reveal the interaction of the insulin-like peptide 3 hormone with the luteinizing hormone in the Leydig cell. The authors reported that the response of MA-10 cells to gonadotropin stimulation by increasing transcription of nur77 and Cyp17 was not comprise acute regulation of insl3 mRNA abundance. The results suggest that the abundance of insl3 mRNA in the MA-10 Leydig cell line needs to be addressed structurally and may be associated with multiple pathways other than chorionic gonadotropin/luteinizing hormone or cAMP regulation [55].

The ability of MLTC-1 cells to respond to gonadotropins until progesterone production in steroidogenesis is like primary Leydig cells. This condition gives advantage to the use of MLTC-1 cells in studies investigating the steroid production. The amount of progesterone, the primary steroid produced by hCG-stimulated MLTC-1 cells, was measured to reveal the effects of diphenyl ether on steroidogenic activity of Leydig cells [56]. It was found that diphenyl ether administered at concentrations that did not affect cell viability did not alter Star expression, but decreased Cyp11a1 and 3β-hsd expression. The authors reported that the decrease in the amount of progesterone was independent of the cAMP pathway. Because of the presence of metabolic differences between mouse and human Leydig cells, they considered this study to be guiding for *in vivo* experiments. [56]. MLTC-1 cells were also used for demonstrated biphasic effects of perfluorooctanoic acid on steroidogenesis [57]. In this study Star and 3β-hsd expressions were significantly reduced with perfluorooctanoic acid treatment in hCG induced MLTC-1 cells whereas no significant changes were recorded for Cyp11a1, 17β-hsd [57]. In another study, MLTC-1 cells were used as models to reveal the interaction of melatonin and testosterone hormones. The knockdown of melatonin receptors in hCG-stimulated MLTC-1 cells resulted in a significant decrease in testosterone levels. They supported that this decrease in testosterone synthesis was associated with the decline in expression levels of StAR, Cyp11a1, Cyp17a1 detected by gene and protein analyses [58].

CONCLUSION

Analysing gonadal testosterone production with immortalized cell lines established to date has been inadequate to give definitive results. Studies have reported that steroidogenesis occurs in cell lines, however, usually results in high amounts of progesterone production, and testosterone production at very low or absent levels. The limitations of Leydig cell lines, such as the inactive enzymes in the basal condition and their inability to maintain stability during long passages, often necessitate confirmation by animal experiments. In addition, all established cell lines originate from mouse or rat testes, causing problems in adapting the results of studies to humans because there are metabolic differences between murine and human Leydig cells. For these reasons, the need for standard Leydig cell line development, such as the H295R human cell line developed for adrenal steroidogenesis, remains important.

REFERENCES

[1] O'Donnell, L., Stanton, P., and de Kretser, David, M. 2017. "Endocrinology of the male reproductive system and spermatogenesis." In *Endotext [Internet],* edited by Feingold, K., R., Anawalt, B., and Boyce. A. South Dartmouth: MDText.com, Inc.

[2] Gnessi L., Fabbri A., and Spera G., 1997. "Gonadal Peptides as Mediators of Development and Functional Control of the Testis: an integrated system with hormones and local environment." *Endocrine Reviews*, 18(4): 541-609 Accessed August 1, 1997. doi: 10.1210/edrv.18.4.0310.

[3] Gartner, L. P and Hiatt, J. L. 2015. *Cell biology and histology.* Maryland: BRS.

[4] Ross, Micheal H. and Pawlina, W. 2003. *Histology: A Text and Atlas,* Philadelphia: Lippincott Williams-Wilkins.

[5] Mescher, Anthony L. 2015. *Junqueira's Basic Histology Text and Atlas,* McGraw Hill Medical Books.

[6] Leydig F. 1850. "On the anatomy of male reproductive organs and anal glands of mammals." *Journal of Scientific Zoology* 2:1–57.

[7] Bouin, P, and Ancel, P. 1903. "Interstitial cell research of the mammalian testis." *Archives of Experimental and General Zoology A*, 1:437–523.

[8] Bouin, P, and Ancel, P. 1904. "Research on the physiological significance of the interstitial gland of mammalian testis. I. Role of the interstitial gland in adult individuals." *Journal of Physiology and Pathology*, 6:1012–22.

[9] Ancel, P., and Bouin, P. 1904. "Researches on the physiological significance of the interstitial gland of mammalian testis. II. Role of the interstitial gland in the embryo, young and old subjects; its functional variations." *Journal of Physiology and Pathology*, 6:1039–50.

[10] Bouin, P., and Ancel P. 1905. "The interstitial gland of the testicle in the horse." *Archives of Experimental and General Zoology A,* 3:391–433.

[11] Wattenberg, Lee, W. 1958. "Microscopic histochemical demonstration of steroid-3β-ol dehydrogenase in tissue sections." *Journal of Histochemistry and Cytochemistry,* 6:225–32. Accessed July 1, 1958. doi: doi.org/10.1177/6.4.225.

[12] Christensen, Kent, A., and Mason, Norman, R. 1965. "Comparative ability of seminiferous tubules and interstitial tissue of rat testes to synthesize androgens from progesterone-4-14C *in vitro*." *Endocrinology* 76:646–56. Accessed April 1, 1965. doi: 10.1210/ endo-76-4-646.

[13] Vandermark, N. L, and Ewing, L. L. 1963. "Factors affecting testicular metabolism and function." *Reproduction*, 6(1):1–8.

[14] Chubb C., and Ewing Larry, L. 1979. "Steroid secretion by *in vitro* perfused testes: Inhibitors of testosterone biosynthesis." *American Journal of Physiology and Endocrinology,* 237(3):239–46. Accessed September 1, 1979. doi: 10.1152/ajpendo.1979.237.3. E239.

[15] Hall P. F, Irby D. C, and De Kretser D. M. 1969 "Conversion of cholesterol to androgens by rat testes: comparison of interstitial cells and seminiferous tubules." *Endocrinology,* 84(3):488–496. Accessed March 1, 1969. doi: 10.1210/endo-84-3-488.

[16] Christensen, A. Kent. 2007. "History of Leydig cell Research." In *Contemporary Endocrinology: The Leydig Cell in Health and*

Disease, edited by Payne H. Anita, Hardy P. Matthew, 3-33. New Jersey: Humana Press.

[17] Svechnikov, K., Izzo G., Andreh L., Weisser J., and Söder O. 2010. "Endocrine Disruptors and Leydig cell Function." *Journal of Biomedicine and Biotechnology*, 684504 Accessed June 23, 2010. doi: 10.1155/2010/684504.

[18] Martin, Luc J., 2016. "Cell interactions and genetic regulation that contribute to testicular Leydig cell development and differentiation." *Molecular Reproduction and Development,* 83(6):470-87. Accessed April 14, 2016. doi: 10.1002/mrd.22648.

[19] Zirkin, Barry, R., and Papadopoulos, Vassilios. 2018. "Leydig cells: formation, function, and regulation." *Biology of Reproduction*, 99(1):101-11. Accessed March 16, 2018. doi: 10.1093/biolre/ioy059.

[20] Rahman, Nafis, A., and Huhtaniemi, Ilpo T. 2004. "Testicular cell lines." *Molecular and Cellular Endocrinology,* 228(1-2):53-65. Accessed July 28, 2004. doi: 10.1016/j.mce.2003.05.001.

[21] Haider, Syed G., 2004. "Cell biology of Leydig cells in the testis." In *International Review of Cytology,* edited by Kwang W. Jeon, 181-241. Tennessee: Elsevier.

[22] Trainer, T. D., 1995. "Testes and Excretory Duct System." In *Histology for Pathologists*, edited by Mills, S. E., 943-62. Philadelphia: Raven Press.

[23] Payne, Anita, H., Youngblood, and Geri, L. 1995. "Regulation of Expression of Steroidogenic Enzymes in Leydig Cells." *Biology of Reproduction*, 52:217-25. Accessed February 1, 1995. doi: 10.1095/biolreprod52.2.217.

[24] Wilson Jean, D. 2001. "Prospects for research for disorders of the endocrine system." *JAMA: The Journal of The American Medical Association*, 285(5): 624-27. Accessed February 7, 2001. doi: 10.1001/jama.285.5.624.

[25] Hardy P. Matthew, Chen G., and Ge R., 2008. "The Role of Leydig cell in spermatogenic function." In *Molecular Mechanisms in Spermatogenesis*, edited by Cheng C. Y, 255-69 New York: Landes Bioscience and Springer Science Business Media.

[26] Stocco Douglass, M. 2000. "Intramitochondrial cholesterol transfer." *Biochimica et Biophysica Acta (BBA) - Bioenergetics*, 1486(1):184-97 Accessed June 8, 2000. doi: 10.1016/S1388-1981(00)00056-1.

[27] Hales, Buchanan, D. 2002 "Testicular macrophage modulation of Leydig cell steroidogenesis." *Journal of Reproductive Immunology*, 57:3–18. Accessed April 24, 2002. doi: 10.1016/S0165-0378(02)00020-7.

[28] Irfan Maqsood, M., Matin, Maryami M., Bahrami, Ahmad R., and Ghasroldasht, Mohammed, M. 2013. "Immortality of cell lines: challenges and advantages of establishment." *Cell Biology International*, 37(10):1038-45. Accessed May 30, 2013. doi: 10. 1002/cbin.10137.

[29] Engeli, Rogel. T., Fürstenberger, C., Kratschmar, Denise V., and Odermatt, A. 2018. "Currently available murine Leydig cell lines can be applied to study early steps of steroidogenesis but not testosterone synthesis." *Heliyon*, 4(2): e00527. Accessed February 1, 2018. doi: 10.1016/j.heliyon. 2018.e00527.

[30] OECD, OECD Guideline for the Testing of Chemicals. Test No.456:H295R Steroidogenesis Assay, OECD Guidelines for the Testing of Chemicals, Section 4: Health Effects, 2011.

[31] Odermatt, A., Strajhar, P., Engeli, and Rogel, T. 2016. "Disruption of steroidogenesis: cell models for mechanistic investigations and as screening tools." *The Journal of Steroid Biochemistry and Molecular Biology*, 158:9-21. Accessed January 22, 2016. doi: 10.1016/j.jsbmb. 2016.01.009.

[32] Shin S, Yasumura Y, and Sato Gordon, H. 1968. "Studies on interstitial cells in tissue culture. II. Steroid biosynthesis by a clonal line of rat testicular interstitial cells." *Endocrinology,* 82:614–16. Accessed March 1, 1968. doi: 10.1210/endo-82-3-614.

[33] Stocco, Douglass, M., and Chen, W. 1991. "Presence of identical mitochondrial proteins in unstimulated constitutive steroid-producing R2C rat Leydig tumor and stimulated nonconstitutive steroid-producing MA-10 mouse Leydig tumor cells." *Endocrinology,*

128:1918–26. Accessed April 1, 1991. doi: 10.1210/endo-128-4-1918.

[34] Ascoli, M., 1981. "Characterization of several clonal lines of cultured Leydig tumor cells: gonadotropin receptors and steroidogenic responses." *Endocrinology,* 108:88–95. Accessed January 1, 1981. doi: 10.1210/endo-108-1-88.

[35] Ascoli, M., 1981. "Regulation of gonadotropin receptors and gonadotropin responses in a clonal strain of Leydig tumor cells by epidermal growth factor." *Journal of Biological Chemistry,* 256:179-83. Accessed January 10, 1981.

[36] Rao, Rekha, M., Jo, Y., Babb-Tarbox, M., Syapin, Peter, J., and Stocco, Douglass, M. 2002. "Regulation of steroid hormone biosynthesis in R2C and MA-10 Leydig tumor cells: role of the cholesterol transfer proteins StAR and PBR." *Endocrine Research,* 28: 387–94. Accessed July 07, 2009. doi:10.1081/ERC-120016813.

[37] Rao, Rekha, M., Jo, Y., Leers-Sucheta, S., Bose, Himangshu, S., Miller, Walter, L., Azhar, S., Stocco, Douglass, M. 2003. "Differential regulation of steroid hormone biosynthesis in R2C and MA-10 Leydig tumor cells: role of SR-B1-mediated selective cholesteryl ester transport." *Biology of Reproduction,* 68:114–21. Accessed January 01, 2003. doi:10.1095/biolreprod.102.007518.

[38] Rebois, R. V. 1982. "Establishment of gonadotropin-responsive murine Leydig tumor cell line." *The Journal of Cell Biology*, 94(1):70-6. Accessed July 1, 1982. doi:10.1083/jcb.94.1.70.

[39] Finaz, C., Lefèvre, A., and Dampfhoffer, D. 1987. "Construction of a Leydig cell line synthesizing testosterone under gonadotropin stimulation: a complex endocrine function immortalized by cell hybridization." *Proceedings of the National Academy of Sciences*, 84(16):5750-753. Accessed August 1, 1987. doi:0.1073/pnas.84.16.5750.

[40] Lefevre, A., Rogier, E., Astraudo, C., Duquenne, C., and Finaz, C. 1994. "Regulation by retinoids of luteinizing hormone/chorionic gonadotropin receptor, cholesterol side-chain cleavage cytochrome P-450, 3 beta-hydroxysteroid dehydrogenase/delta (5-4)-isomerase

and 17 alpha-hydroxylase/C17-20 lyase cytochrome P-450 messenger ribonucleic acid levels in the K9 mouse Leydig cell line." *Molecular and Cellular Endocrinology*, 106:31–39. Accessed January 13, 2003. doi:10.1016/0303-7207(94)90183-X.

[41] Steinberger, E., Steinberger, A., and Ficher, M. 1970. "Study of spermatogenesis and steroid metabolism in cultures of mammalian testes." *Recent Progress in Hormone Research,* 26:547–88. Accessed October 21, 2013. doi:10.1016/B978-0-12-571126-5.50018-2.

[42] Ascoli, M. 2007. "Immortalized Leydig cell lines as models for studying Leydig cell physiology." In *The Leydig cell in health and disease,* edited by Payne H. Anita, Hardy P. Matthew, 373-381. New Jersey: Humana Press.

[43] Mather, Jennie P., 1980. "Establishment and characterization of two distinct mouse testicular epithelial cell lines." *Biology of Reproduction*, 23:243–52. Accessed August 1, 1980. doi:/10.1095/biolreprod23.1.243.

[44] Ohta, S., Tabuchi, Y., Yanai, N., Asano, S., Fuse, H., and Obinata, M. 2002. "Establishment of Leydig cell line, TTE1, from transgenic mice harboring temperature-sensitive simian virus 40 large T-antigen gene." *Archives of Andrology*, 48(1):43-51. Accessed July 09, 2009. doi:10.1080/014850102753385206.

[45] Ohta, S., Fuse, H., and Tabuchi, Y. 2002. "DNA microarray analysis of genes involved in the process of differentiation in mouse Leydig cell line TTE1." *Archives of Andrology*, 48(3):203-08. Accessed July 09, 2009. doi:10.1080/01485010252869298.

[46] Kananen, K., Markkula, M., El-Hefnawy, T., Zhang, Fu, P., Paukku, T., Su, Jyan Gwo, Hsueh, Aaron, J. W., J., and Huhtaniemi, I. 1996. "The mouse inhibin α-subunit promoter directs SV40 T-antigen to Leydig cells in transgenic mice." *Molecular and Cellular Endocrinology*. 119(2):135-46. Accessed March 8, 1999. doi:10.1016/0303-7207(96)03802-6.

[47] Stalvey, John R. D., and Anita, Payne H. 1983. "Luteinizing hormone receptors and testosterone production in whole testes and purified Leydig cells from the mouse: differences among inbred

strains." *Endocrinology* 112:1969-701. Accessed May 1, 1983. doi:10.1210/endo-112-5-1696.

[48] Forgacs, Agnes, L., Ding, Q., Jaremba, Rosemary. G., Huhtaniemi, Ilpo, T., Rahman, Nafis A., Zacharewski, and Timothy R. 2012 "BLTK1 murine Leydig cells: a novel steroidogenic model for evaluating the effects of reproductive and developmental toxicants." *Toxicological Sciences*, 127(2):391-402. Accessed March 29, 2012. doi:10.1093/toxsci/kfs121.

[49] Jørgensen, N., Joensen, U. N., Toppari, J., Punab, M., Erenpreiss, J., Zilaitiene, B., Paasch, U., Salzbrunn, A., Fernandez, M., F., Virtanen H., E., Matulevicius, V., Olea, N., Jensen, T., K., Petersen, J., H., Skakkebæk, N., E., and Andersson, A., M., 2016. "Compensated reduction in Leydig cell function is associated with lower semen quality variables: a study of 8182 European young men." *Human Reproduction*, 31(5):947-57. Accessed March 02, 2016. doi:10.1093/humrep/dew021.

[50] Goncalves, Dutra, G., Semprebon, Cristine, S., Biazi, Isabela, B., Mantovani, Sergio, M., and Fernandes, Alves, Scantamburlo, A. 2018. "Bisphenol A reduces testosterone production in TM3 Leydig cells independently of its effects on cell death and mitochondrial membrane potential." *Reproductive Toxicology,* 76: 26-34. Accessed December 13, 2017. doi: 10.1016/j.reprotox.2017.12.002.

[51] Banerjee, B., Chakraborty, S., Chakraborty, P., Ghosh, D., and Jana, K. 2019. "Protective effect of resveratrol on benzo (a) pyrene induced dysfunctions of steroidogenesis and steroidogenic acute regulatory gene expression in Leydig cells." *Frontiers in Endocrinology*, 10:272. Accessed April 30, 2019. doi: 10.3389/fendo.2019.00272.

[52] Yilmaz, Banu O., Korkut, A., and Erkan, M. 2018. "Sodium fluoride disrupts testosterone biosynthesis by affecting the steroidogenic pathway in TM3 Leydig cells." *Chemosphere*, *212*:447-55. Accessed 22, 2018. doi: 10.1016/j.chemosphere.2018.08.112.

[53] Gao, Y., Wu, X., Zhao, S., Zhang, Y., Ma, H., Yang, Z., Yang, W., Zhao, C., Wang, L., and Zhang, Q. 2019. "Melatonin receptor

depletion suppressed hCG-induced testosterone expression in mouse Leydig cells." *Cellular & Molecular Biology Letters*, 24(1):21. Accessed March 11, 2019. doi:10.1186/s11658-019-0147-z.

[54] Eze, Ukpai A., Huntriss, John D., Routledge, Michael N., and Gong, Yun Y. 2019. "*In vitro* effects of single and binary mixtures of regulated mycotoxins and persistent organochloride pesticides on steroid hormone production in MA-10 Leydig cell line." *Toxicology in vitro* 60:272-80. Accessed June 11, 2019. doi: 10.1016/j.tiv.2019. 06.007.

[55] Strong, Mary E., Matthew, Burd A., and Daniel, Peterson G. 2019. "Evaluation of the MA-10 cell line as a model of insl3 regulation and Leydig cell function." *Animal Reproduction Science* 208: 106116. Accessed June 28, 2019. doi: 10.1016/j.anireprosci.2019.106116.

[56] Han, X., Wang, Y., Chen, T., Wilson, Mark J., Pan, F., Wu, X., Rui, C., Chen, Tang, Q., and Wu, W. 2019. "Inhibition of progesterone biosynthesis induced by deca-brominated diphenyl ether (BDE-209) in mouse Leydig tumor cell (MLTC-1)." *Toxicology in vitro*, 60:383-88. Accessed May 24, 2019. doi: 10.1016/j.tiv.2019.05.016.

[57] Tian, M., Huang, Q., Wang, H., Martin, Francis L., Liu, L., Zhang, J., and Shen, H. 2019. "Biphasic effects of perfluorooctanoic acid on steroidogenesis in mouse Leydig tumor cells." *Reproductive Toxicology*, 83:54-62. Accessed November 30, 2018. doi: /10.1016/ j. reprotox.2018.11.006.

[58] Gao, S., Chen, S., Chen, L., Zhao, Y., Sun, L., Cao, M., Huang, Y., Niu, Q., Wang, F., Yuan, C., Li, C., and Zhou, X. 2019. "Brain-derived neurotrophic factor: A steroidogenic regulator of Leydig cells." *Journal of Cellular Physiology*, 234(8):14058-67. Accessed January 9, 2019. doi:10.1002/jcp.28095.

Biographical Sketch

Nebahat Yildizbayrak

Affiliation: Istanbul University, Science Faculty, Department of Biology, Istanbul/Turkey.

Education: Istanbul University, Science Faculty, Department of Biology, PhD.

Business Address: Istanbul University, Science Faculty, Department of Biology, 34134, Vezneciler/ Istanbul/ Turkey.

Research and Professional Experience: Cell Biology, Reproductive Biology, Cytotoxicity, Genotoxicity, Toxicology.

Publications from the Last 3 Years:

1. Orta-Yilmaz B., Yildizbayrak, N., Aydin, Y., and Erkan, M. 2017. "Evidence of acrylamide and glycidamide induced oxidative stress and apoptosis in Leydig and Sertoli cells" *Human and Experimental Toxicology*, Accessed January 8, 2017. doi:10.1177/096032711 6686818.
2. Yildizbayrak, N., and Erkan, Melike, B. 2018. "Acrylamide disrupts the steroidogenic pathway in Leydig cells: possible mechanism of action." *Toxicological and Environmental Chemistry*, Accessed April 20, 2018. doi: 10.1080/02772248.2018.1458231.
3. Orta-Yilmaz, B., Yildizbayrak, N., and Erkan, M. 2018. "Sodium arsenite-induced detriment of cell function in Leydig and Sertoli cells: the potential relation of oxidative damage and antioxidant defense system." *Drug and Chemical Toxicology*, Accessed September 12, 2018. doi: 10.1080/01480545.2018.1505902.
4. Perker, Mehmet, C., Orta-Yilmaz, B., Yildizbayrak N., Aydin Y., and Erkan, Melike, B. 2019. "Protective effects of curcumin on

biochemical and molecular changes in sodium arsenite induced oxidative damage in embryonic fibroblast cells." *Journal of Biochemical and Molecular Toxicology*, Accessed April 1, 2019. doi:10.1002/jbt.22320.

5. Yildizbayrak, N., and Erkan, Melike, B. 2019. "Therapeutic effect of curcumin on acrylamide induced apoptosis mediated by MAPK signaling pathway in Leydig cells." *Journal of Biochemical and Molecular Toxicology*, Accessed May 13, 2019. doi:10.1002/jbt. 22326.
6. Aydin, Y, Orta, Yilmaz, B., Yildizbayrak, N., Korkut, A., Arabul, Kursun, M., Irez, T., and Erkan, Melike, B. 2019. "Evaluation of citrinin-induced toxic effects on mouse Sertoli cells." *Drug and Chemical Toxicology*, Accessed May 29, 2019. doi:10.1080/0148 0545.2019.1614021.

In: Leydig Cells
Editor: Bruno Solomon

ISBN: 978-1-53617-282-9

Chapter 3

THE ROLE OF ENVIRONMENTAL CONTAMINANTS IN LEYDIG CELL STEROIDOGENESIS

Banu Orta Yilmaz** and Melike Erkan
Department of Biology, Faculty of Science,
Istanbul University, Istanbul, Turkey

ABSTRACT

In the world, population growth, dietary habits, industrialization and the progress of agriculture have increased the frequency of exposure of people to various pollutants. Recent studies have shown that environmental pollutants cause many damages on male reproductive health. Environmental pollutants lead to male infertility, including the development of testicular cancer, decreased sperm count, impaired semen quality, deteriorated spermatogenesis and decreased testosterone production in the male reproductive system. It has been found that many environmental chemicals target endocrine system parameters that play a role in the reproductive function and cause various reproductive abnormalities. However, the basic mechanisms and molecular pathways

* Corresponding Author's Email: banu.yilmaz@istanbul.edu.tr.

involved in these anomalies have not been fully elucidated. In this context, it is important to detect changes in cell function of Leydig cells responsible for testosterone biosynthesis, which play a primary role in male reproductive system. This review discusses the effects of the most commonly used environmental contaminants on Leydig cell function through the impaired of steroidogenic pathway.

Keywords: Leydig cell, steroidogenesis, chemical pollutants, testosterone biosynthesis

ABBREVIATIONS

17β-HSD	17β-Hydroxysteroid Dehydrogenase
3β-HSD	3β-Hydroxysteroid Dehydrogenase
Arom	Aromatase
BPA	Bisphenol A
cAMP	Cyclic Adenosine Monophosphate
Cyp17a1	17α-Hydroxylase / C17-20 Lyase
DBP	Dibutyl Phthalate
hCG	Human Chorionic Gonadotropin
LH	Luteinizing Hormone
LHR	Luteinizing Hormone Receptor
MBP	Monobutyl Phthalate
mLTC-1	Murine Leydig Tumour Cell Line 1
P450scc	Cytochrome P450
PCBs	Polychlorinated Biphenyls
PRL	Prolactin
PVC	Polyvinyl Chloride
SR-B1	Scavenger Receptor Class B Type I
StAR	Steroidogenic Acute Regulatory Protein
T3	Triiodothyronine
T4	Thyroxine
TSH	Thyroid-Stimulating Hormone.

Introduction

The increase in population and industrialization day by day caused the harmful chemicals to enter the environment. Humans are exposed to these environmental toxicants from contaminated food and water, cosmetics, industrial products, air and cigarettes [1, 2]. It is known that the accumulation of these toxic components leads to many diseases threatening human health, such as haematological, hepatic, renal and neurological diseases. With increasing infertility, recent studies have focused on investigating the possible effects of environmental contaminants on the male reproductive system [3, 4]. Reports have shown that hormonal disorders, drugs used, diseases of the reproductive organs, testicular cancer are the leading causes of male infertility [5, 6]. Many environmental pollutants, due to their endocrine disrupting effects, cause suppression of reproductive capacity, decrease in testicular spermatozoa concentration and testicular weight. Besides, these chemicals cause a decrease in the number of Leydig and Sertoli cells, inhibit testosterone synthesis and the activity of steroidogenic enzymes in Leydig cells and block to spermatogenesis [7, 8]. The number of various environmental contaminants examined for their effects on reproductive health is quite high. These include a wide variety of elements and classes of compounds such as heavy metals, organic polychlorinated dibenzodioxins, polychlorinated biphenyls, phthalates, organochlorine pesticides, dicarboximide fungicides, phenols and food contaminants in the preparation process (Figure 1). Although the use of certain toxic substances such as polychlorinated biphenyls is prohibited in many countries, it is known that these substances do not disappear in nature and continue to be exposed [9]. However, although many chemicals, such as heavy metals, pesticides, have been imposed limits by organizations such as World Health Organization, Environmental Protection Agency, International Agency for Research on Cancer, they can still be detected at significant concentrations in the environment. This chapter focuses on various chemical contaminants that inhibit Leydig cell steroidogenesis and the pathways that these chemicals affect during the steroidogenesis process.

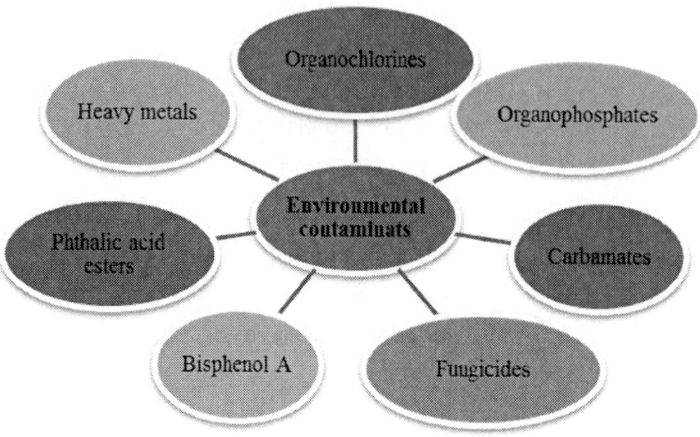

Figure 1. Environmental contaminants.

Literature Search

In online databases (PubMed, ScienceDirect, and Google Scholar), scanned reference lists and journals, potentially suitable studies have been searched to investigate the role of environmental contaminants on Leydig cell steroidogenesis. The evaluated studies included research articles in English language which can only be accessed in the full text and the last research date was August 25, 2019. A total of 63 papers were selected to be included in this review.

Leydig Cells

The male reproductive system is responsible for producing sperm and regulating reproductive function with various hormones [10]. The male reproductive system consists mainly of testes, epididymis, vas deferens, ejaculation channels, urethra, seminal vesicle, prostate gland, bulbourethral glands and penis. While the testes are mainly responsible to produce sperm and male sex hormones, the prostate, bulbourethral glands and vesicular seminalis, which are the auxiliary sex glands, produce seminal fluid that is excreted in the ejaculation with sperm. [11].

The testis consists of two important compartments: interstitial cells and seminiferous tubules. The lumen of the seminiferous tubules where the spermatogenesis process takes place is lined with epithelial layer and consists of spermatogenetic cells (spermatogonium, spermatocytes, spermatids), multifunctional Sertoli cells and myoid cells surrounding the tubules. [12, 13]. Unlike spermatogenetic cells, Sertoli cells are prismatic cells extending from the basal lamina to the seminiferous tubule lumen and provide mechanical support to the spermatogenic cells with these positions. After puberty, the lumen of the seminiferous tubules is divided into two parts, basal and ad luminal, through tight junction in the basal parts of Sertoli cells. These tight junction complexes are responsible for forming the blood-testicular barrier that protects developing spermatogenic cells from the autoimmune reaction [14, 15].

The interstitial area is located between the seminiferous tubules. In this area, nerve fibers, blood and lymph vessels, mast cells, fibroblasts, macrophages and Leydig cells are found. Leydig cells are endocrine cells that produce and secrete testosterone through the luteinizing hormone secreted from the pituitary. Leydig cells are responsible for the development of secondary sex characteristics such as sound thickening, male-specific muscle structure and beard and moustache development by the testosterone hormone they secrete. The secreted testosterone hormone is transported by blood and affects the functions of the prostate, bulbourethral glands and vesicular seminalis supporting the male reproductive system. Due to its ability to produce steroids, Leydig cells have lipid droplets, crystalline mitochondria and a well-developed smooth endoplasmic reticulum [10].

Leydig cells in mammals contain two different types. These are fetal type Leydig cells responsible for the formation of male primary sex characters during fetal and neonatal period and adult type Leydig cells responsible for pubertal male characteristics [16]. Fetal Leydig cells produce androgens that are essential for male fetus masculinization at the embryonic period. During the early postnatal period, adult Leydig cells replace fetal Leydig cells. Adult Leydig cells produce the necessary testosterone for the completion of sexual differentiation and reproductive functions in men [17]. Among the main functions of testosterone secreted by Leydig cells are the occurrence of sexual desire (libido), maintenance of genital glands, the occurrence of secondary sex characteristics, control of spermatogenesis, negative feedback in the pituitary and hypothalamus [18]. Leydig cell function is regulated by two hormones of the anterior pituitary: luteinizing hormone (LH), while stimulating testosterone production; prolactin initiates LH receptor expression. The synthesis and secretion of testosterone from Leydig cells occurs by LH stimulation. LH is the main factor for the differentiation and maintenance of Leydig cells [19].

Testosterone production of Leydig cells decreases in increased age. Although this mechanism is not fully elucidated, possible causes include a decrease in Leydig cell count, a decrease in LH levels, and a decrease in capacity of Leydig cells to produce testosterone [20].

Leydig Cell Steroidogenesis

The function of Leydig cells is to perform testosterone biosynthesis, which is necessary for male sexual differentiation, maintenance of spermatogenesis and expression of male secondary sex characteristics [18]. Approximately 95% of testosterone in the serum is synthesized by Leydig cells and 5% by the adrenal cortex. Acute or chronic testosterone biosynthesis is due to the stimulation of Leydig cells by LH which is a pituitary hormone [21]. The LH binds to the specialized high affinity receptor on the surface of Leydig cells. Binding of LH to its receptor results in activation of adenylate cyclase followed by increased production of the intracellular secondary messenger cAMP. Acute and chronic effects of LH are associated with an increase in cAMP [22]. Various factors, such as cell skeleton elements and steroid transporter protein, are involved in the transfer of free intracellular cholesterol to the outer mitochondrial membrane. Cholesterol is stored for use in lipid droplets that are not enveloped by the membrane. The acute response is rapid and involves the transport of cholesterol to the internal mitochondrial membrane, the site of the first enzyme on the path from cholesterol to testosterone. There are two carrier proteins in the transport of cholesterol from the outer membrane of the mitochondria to the inner membrane: 1- StAR, 2- Peripheral-Type Benzodiazepine Receptor [23].

Chronic stimulation of Leydig cells by LH or cAMP requires optimal expression of the enzymes necessary for the biosynthesis of testosterone from cholesterol. The first step of this pathway is the conversion of C27 cholesterol to C21 steroid, pregnenolone. This reaction is catalysed by the cholesterol side chain cleavage enzyme (P450scc) which is the cytochrome P450 enzyme. Pregnenolone spreads to mitochondrial membranes and is metabolized by enzymes associated with the smooth endoplasmic reticulum. Pregnenolone passes through the smooth endoplasmic reticulum, where 3β-hydroxy dehydrogenase (3β-HSD) will produce progesterone. Progesterone is then converted to 17α-hydroxy progesterone and then to androstenedione via the enzyme 17α-hydroxylase / C17-20 lyase (Cyp17a1). In the last step, androstenedione forms testosterone with the effect of 17β-hydroxysteroid dehydrogenase (17β-HSD) enzyme [22].

Enzymes involved in testosterone biosynthesis can be divided into two major classes of proteins: cytochrome P450 heme-containing proteins; CYP11A1 (P450scc), CYP17A1 (P450c17) and P450arom and hydroxysteroid dehydrogenases; 3β-hydroxysteroid dehydrogenase (3β-HSD) and 17β-hydroxysteroid Dehydrogenase (17β-HSD) [24].

The mechanisms controlling gene expression are the basis of many biological processes in health and disease. In order to solve these mechanisms, it is necessary to understand the transcription factors that direct tissue specific and hormonal regulated gene expression. Transcription factors are nuclear proteins that regulate tissue-specific gene expression by acting as positive or negative regulators of gene expression [25]. Approximately 3000 transcription factors have been identified in eukaryotes and classified based on similarities in amino acid sequence of DNA binding domains [16]. Transcription factors are studied in four main domains: the domains, the domains with the zinc-dependent DNA binding region, the helical-spin-helical and β-scaffold.

Testicular formation, Leydig cell development and function are based on a transcriptional network encoded in DNA. This process requires the interaction of various factors, such as signalling molecules, receptors and transcription factors. Several transcription factors from various families have been associated with testicular formation, Leydig cell development and function. Most of the transcription factors known to be expressed in Leydig cells have been identified by analysis of regulatory regions of genes expressed in these cells, and in particular by analysis of genes encoding steroidogenic enzymes involved in testosterone synthesis [25]. Sf-1, Gata-4 and Nur77 transcription factors are the major transcription factors that encode multiple genes in the steroidogenic pathway for testosterone biosynthesis [26].

ENVIRONMENTAL CONTAMINANTS

Many chemicals in the environment are essential for human life. The change in nutritional habits has increased food contaminants, with the development of industry there has been an increase in waste and air pollution, making these beneficial compounds harmful to human health. Although these environmental pollutants are classified in various ways, compounds such as dioxins, phthalates, polychlorinated biphenyls, pesticides, heavy metals, bisphenol a, and methoxychlor can be grouped as toxins that people are most exposed to daily. These toxic compounds are generally pass into the air, water and nutrients and cause many losses of function in the respiratory, excretory, digestive and reproductive systems in humans. Some of these toxic compounds may cause serious problems on the reproductive system related to infertility and may even be transferred onto the next generations. Determination of these reproductive toxicants and revealing the underlying mechanisms are very important. This section includes the effects of the most common reproductive toxicants on Leydig cell steroidogenesis.

Bisphenol A

Bisphenol A (2,2-bis [4-hydroxyphenyl] propane, BPA) is an endocrine disrupter widely used in the world and unavoidable for living organisms [27]. BPA is used in a wide range such as disposable cups, plastic bottles, perfumes, food containers, toothpastes. BPA poses a risk to human health, from cancer, diabetes, obesity, heart disease to behavioural disorders. Many studies have shown that BPA also lead to toxic effects on male reproductive system and causes infertility due to its endocrine disruptive properties [28]. In a previous study, it was found that a concentration of 0.01 nM BPA suppressed testosterone production in mouse Leydig cells, which could be mediated by inhibition of the gene encoding the cytochrome P450 17α-hydroxylase / 17,20 lyase enzyme [29]. In Leydig cells obtained from rats by primary culture method, BPA has been shown to reduce the levels of luteinizing hormone receptor (LHR) and HSD17B3 that activate testosterone formation [30]. In another study, BPA concentrations were selected as concentrations similar to those found in serum of occupationally exposed individuals and their effects were investigated in Leydig TM3 cells. As a result, it was found that BPA significantly decreased testosterone production and impaired Leydig cell function [31]. In a study conducted with MA-10 Leydig cells, BPA was found to affect CYP19 and CYP11a1 gene expression levels by disrupting testosterone biosynthesis [32]. The effects of BPA on steroidogenesis in the mLTC-1 Leydig cell line were evaluated and it was found that BPA suppressed progesterone production by decreasing the expression of the scavenger receptor class B type I (SR-B1) and cytochrome P450 (P450scc) in hCG-stimulated mLTC-1 cells [33].

Phthalates

Phthalates are commonly used in the industry to make the hard and brittle structure of plastics made of polyvinyl chloride (PVC) soft and flexible. In addition, it can be used in many different areas such as solvent, lubricating oil and cleaner, so that human exposure to phthalates is inevitable. Phthalates are used in shoes, stationery products, toys, cosmetics and textile products, which are used in many fields but must be used continuously in line with daily needs [34-36]. Many animal and human studies have shown that phthalates disrupt steroid synthesis, reduce the amount of testicular testosterone, resulting in dysfunction in male reproduction [37, 38]. Dibutyl phthalate (DBP) is a commonly used synthetic phthalate and monobutyl phthalate (MBP) is its main metabolite [39]. In one study, it was shown that MBP and DBP inhibited androgen production in immature Leydig cells, also decreased the expression levels of Cyp11a1, Star, Hsd3b1 and Hsd17b3 [40]. Low MBF concentrations (20–200 μM) increased the testosterone concentration of adult Leydig cells by inducing StAR, P450scc, 3β-HSD and 17β-HSD enzymes. Contrary to the effect at low concentrations, high levels of MBF (2000 μM and above) inhibited steroidogenesis by reducing the activity of these enzymes in adult Leydig cells [41]. In a study with MA-10 Leydig cells, MBP exposure showed reductions in LH-stimulated cAMP and progesterone levels. According to the findings, MBP has been shown to inhibit MA-10 Leydig cell steroidogenesis by targeting LH-induced cAMP production and cholesterol transport [42].

PCBs

Polychlorinated biphenyls (PCBs) are industrial compounds that do not easily disappear in nature due to their chemical stability. PCBs are mostly used in transformers, hydraulic and vacuum pumps, furniture, paints and printing ink, pesticide additives, adhesives and electrical insulation fluids [43]. Humans are exposed to PCBs by inhalation, contaminated water and food intake and occupational exposure. PCBs show endocrine disruptive properties in animals even at low concentrations [44]. In addition, PCBs has immunosuppressive, neurotoxic, carcinogenic, teratogenic effects and causes behaviour disorder [45, 46]. Endocrine disruptive properties of

PCBs affect male reproductive system and cause infertility [47]. Estrogenic and antiestrogenic PCB congeners and / or metabolites of congeners can be directly linked to the oestrogen receptor and steroid-binding globulin [48]. PCB exposure in the male during development period leads to pathological conditions such as cryptorchidism, testicular cancer and infertility. In a study, PCBs (Aroclor 1254) were significantly reduced in Leydig cellular mRNA and protein expressions of 5α-reductase, aromatase and AR [49]. Administration of PCBs caused a decrease in the Leydig cell population with decreased activity of steroidogenic enzymes in 3β-HSD and 17β-HSD. Moreover, it was determined that PCBs showed a significant reduction in LHR, SR-B1, StAR protein, Cyp11a1, 3β-HSD, Cyp17a1, 17β-HSD, 5α-reductase, Cyp19a1 and AR gene expression in Leydig cells [50]. In a study examining the effects of PCBs (Aroclor 1254) exposure in isolated Leydig cells, it was found that Aroclor 1254 administration significantly reduced LH, TSH, PRL, T3, T4, testosterone and estradiol. In addition, Aroclor 1254 was found to decrease cytochrome P450scc, 3β-HSD, 17β-HSD enzyme levels significantly in Leydig cells [47]. In a study by Aydin and Erkan, Aroclor 1242, a PCB derivative, was found to inhibit steroidogenesis by inhibiting 3β-HSD and 17β-HSD steroidogenic enzyme in Leydig cells [51].

Heavy Metals

Heavy metals are elements that have serious toxic effects on humans, animals and plants, even at low concentrations. Among the heavy metals with the highest toxicity are lead, mercury, cadmium, arsenic and cobalt. Common routes of exposure to these heavy metals are contaminated drinking water, vegetable and animal foods, water pipes, cosmetics, insecticides, dyes and cigarettes. Heavy metals can enter the organism through the mouth, skin and respiration, and accumulation in the body can result in many health problems such as hepatological, neurological, dermatological, cardiac and infertility [52]. Heavy metals affecting reproductive health adversely disrupt fertility by causing abnormal sperm concentration and motility, increased sperm count, suppressed steroidogenic gene expression levels, impaired spermatogenesis and hormonal imbalance [53]. In a study conducted with lead which is one of the heavy metals, it was shown that lead induced a decrease in the level of StAR protein expression and steroidogenic enzymes 3β-hydroxysteroid dehydrogenase (3β-HSD) and CYP11A1 in Leydig cells [54]. In a previous study with lead exposed R2C Leydig cells, it was concluded that lead significantly decreased the progesterone concentration and suppressed expression levels of steroidogenic proteins Star, 3β-HSD and CYP11A1 [55]. In a previous study, cadmium was shown to reduce the levels of CYP11A1, HSD3B1, CYP17A1 and HSD17B3 in fetal Leydig cells and block testosterone biosynthesis. Furthermore, the mechanism underlying the testosterone imbalance was thought to be due to the reduction of Star gene expression [56]. Recent report with arsenic, which is one of the frequently studied heavy metals, showed suppression mRNA expression of the Lhr, Star, P450scc, Hsd3b, Cyp17a1, Hsd17b and Arom genes, which play an important role in the steroidogenic pathway in Leydig cells [57].

PESTICIDES

Pesticides are detrimental chemicals used to control pests such as bacteria and viruses or to eliminate their harmful effects. Pesticides are frequently used in agricultural fields and interfere with the basic biological functions of the target organism. The major types of pesticides are bactericides, insecticides, fungicides, herbicides, and humans are exposed to these different types of pesticides from various sources. Contaminated vegetables and fruit, insecticides, occupational exposure, contaminated water and air can be among the ways of infection to humans. While the duration and level of exposure to pesticides is of great importance, acute intoxications and allergic reactions are common in the short term. Long-term exposure to pesticides may result in loss of function in the circulatory, excretory and reproductive systems, as well as cause cancer related to genetic damage. Also, pesticides have negative effects on male reproductive health due to their endocrine disruptive effects. In a study, acetamiprid, an insecticide, significantly inhibited the expression levels of cAMP-dependent steroidogenic proteins (STAR, HSD3B, CYP11A1) by reducing cAMP levels in Leydig cells. As a result, acetamiprid disrupted testosterone biosynthesis, causing loss of Leydig cell function [58]. In previous report demonstrated that herbicide roundup inhibits steroidogenesis in Leydig cells by disrupting expression of the steroidogenic acute regulatory (StAR) protein [59]. In a study investigating the effects of cypermethrin, which is known to be an insecticide, on Leydig cell steroidogenesis, it stimulated the production of testosterone and the expression of genes associated with steroidogenesis (StAR and CYP11A1), via disrupting the hypothalamic-pituitary-gonadal axis [60]. Moreover, permethrin, an insecticide, exposure significantly depressed the testosterone production and testicular mitochondrial mRNA and protein expression levels of StAR and P450scc [61]. In a study examining the effects of mycotoxins together and separately, MA-10 Leydig cells were exposed to zearalenone, ochratoxin and deoxynivalenol for 48 hours. According to the results of the study, these mycotoxins decreased testosterone and progesterone levels significantly and downregulated the expression levels of the genes responsible for testosterone biosynthesis, thus disrupted Leydig cell steroidogenesis [62]. In a previous study, methoxychlor increased Scarb1 and Cyp11a1 mRNA levels and CYP11A1 protein levels, also significantly decreased testosterone levels in Leydig cells isolated from rats. As a result, it was found that methoxychlor disrupts the balance of Leydig cell steroidogenesis and causes malformations [63].

CONCLUSION

Several studies have clearly demonstrated that environmental contaminants cause an disturbance normal testicular spermatogenesis and steroidogenesis. The investigation of Leydig cell steroidogenesis is crucial in revealing the underlying mechanisms of infertility caused by environmental contaminants. In response to a chemical agent, investigation of changes in the expression levels of molecules involved in steroid production and steroidogenesis in Leydig cells may predict infertility. Such studies give us information about the effects of toxicants on reproductive health, but are not sufficient to prevent advers effects. Finally, the survival of all living organisms in a safe and healthy world depends on limiting the use of existing environmental pollutants and developing new ways to eliminate toxicity for naturally occurring toxic compounds.

REFERENCES

[1] Aitken, John R., Peter Koopman, and Sheena E. M Lewis. 2004. "Seeds of concern." *Nature* 432: 7013 - 48.

[2] Sharpe, Richard M. 2010. "Environmental / lifestyle effects on spermatogenesis." *Philosophical Transactions of the Royal Society B: Biological Sciences* 365. 1546: 1697-1712.

[3] Jenardhanan, Pranitha, Manivel Panneerselvam, and Premendu P. Mathur. 2016. "Effect of environmental contaminants on spermatogenesis." In *Seminars in Cell & Developmental Biology* 59: 126-140.

[4] Knez, Jure. 2013. "Endocrine-disrupting chemicals and male reproductive health." *Reproductive Biomedicine Online* 26.5: 440-448.

[5] Skakkebaek, N. E., Rajpert E. De Meyts, Rajpert E., and K. M. Main. 2001. "Testicular dysgenesis syndrome: an increasingly common developmental disorder with environmental aspects: Opinion." *Human Reproduction* 16.5: 972-978.

[6] Toppari, Jorma, et al. 1996. "Male reproductive health and environmental xenoestrogens." *Environmental Health Perspectives* 104.4: 741–803.

[7] Fox, Glen A. 2001. "Wildlife as sentinels of human health effects in the Great Lakes-St., Lawrence Basin." *Environmental Health Perspectives* 109.6: 853–861.

[8] Oehninger, Sergio. 2001. "Strategies for the infertile man." *Seminars in reproductive medicine* 19: 231–237.

[9] Rudel, Ruthann A., and Laura J. Perovich. 2009. "Endocrine disrupting chemicals in indoor and outdoor air." *Atmospheric Environment* 43: 170–181.

[10] Kierszenbaum, Abraham L., and Laura Tres. 2015. *Histology and Cell Biology: an introduction to pathology E-Book.* Elsevier Health Sciences.

[11] Gartner, Leslie P., and James L. Hiatt. 2001. *Color Textbook of Histology*. Second Edition, W. B.

[12] Bloom, W., and Fawcett, D. W. 1994. *A Textbook of Histology*, ed. 12. Chapman & Hall, New York.

[13] Junquueira, Luis C., and Carnerio, J. 2003. *Basic Histology: Text & Atlas*. Tenth Edition, The McGraw-Hill Companies, Brazil, 0891-2106.

[14] Edwarts, Robert G., and Steven A. Brody. 1995. *Principles and Practice of Assisted Human Reproduction.* W. B. Saunders Company, Pennsylvania, 55-58.

[15] Ross, Michael H., Kaye, G. 1., Pawlina, W. 2003. *Histology: A Text and Atlas*. 4th ed., Lippincott Williams-Wilkins, Philedelphia 696-701.

[16] Haider, Syed G. 2004. "Cell biology of Leydig cells in the testis," *International Review of Cytology* 233, 181–241.

[17] Habert, Rene, Herve Lejeune, and Jose M. Saez. 2001. Origin, differentiation and regulation of fetal and adult Leydig cells. *Molecular and Cellular Endocrinology* 179: 47–74.

[18] Payne, Anita H., and Geri L. Youngblood. 1995. "Regulation of expression of steroidogenic enzymes in Leydig cells." In *Biology of Reproduction* 52: 217-225.

[19] Ge, Ren S., et al. 2005. "Gene expression in rat Leydig cells during development from the progenitor to adult," *Biology of Reproduction* 104: 037499.

[20] Chen, Haolin, et al. 1994. "Age-related decreased Leydig cell testosterone production in the Brown Norway rat." *Journal of Andrology* 15.6: 551-557.

[21] Stocco, Douglas M. 2000. "Intramitochondrial cholesterol transfer." *Biochimica et Biophysica Acta (BBA)-Molecular and Cell Biology of Lipids* 1486.1: 184-197.

[22] Hales, Dale B. 2002. "Testicular macrophage modulation of Leydig cell steroidogenesis." *Journal of Reproductive Immunology* 57.1-2: 3-18.

[23] Stocco, Douglas M. 2014. "The role of PBR/TSPO in steroid biosynthesis challenged." *Endocrinology* 155.1 1: 6–9.

[24] Sanderson, J. Thomas. 2006. "The steroid hormone biosynthesis pathway as a target for endocrine-disrupting chemicals." *Toxicological Sciences* 94.1: 3-21.

[25] Tremblay, Jacques J. 2007. "Transcription factors as regulators of gene expression during Leydig cell differentiation and function." *The Leydig Cell in Health and Disease.* Humana Press, 333-343.

[26] King, Steven R., and Holly A. LaVoie. 2012. "Gonadal transactivation of STARD1, CYP11A1 and HSD3B." *Front Biosci* 17.597: 824-846.

[27] Klinge, Carolyn M. 2000. "Estrogen receptor interaction with co-activators and co-repressors." *Steroids* 65.5: 227-251.

[28] Jin, Pengpeng, et al. 2013. "Low dose bisphenol A impairs spermatogenesis by suppressing reproductive hormone production and promoting germ cell apoptosis in adult rats." *Journal of Biomedical Research* 27.2: 135.

[29] Akingbemi, Benson T., et al. 2004. "Inhibition of testicular steroidogenesis by the xenoestrogen bisphenol A is associated with reduced pituitary luteinizing hormone secretion and decreased steroidogenic enzyme gene expression in rat Leydig cells." *Endocrinology* 145.2: 592-603.

[30] Nanjappa, Manjunatha K., Liz Simon, and Benson T. Akingbemi. 2012. "The industrial chemical bisphenol A (BPA) interferes with

proliferative activity and development of steroidogenic capacity in rat Leydig cells." *Biology of Reproduction* 86.5: 135-1.

[31] Goncalves, Gessica D., et al. 2018. "Bisphenol A reduces testosterone production in TM3 Leydig cells independently of its effects on cell death and mitochondrial membrane potential." *Reproductive Toxicology* 76: 26-34.

[32] Lan, Hsin C., et al. 2017. "Bisphenol A disrupts steroidogenesis and induces a sex hormone imbalance through c-Jun phosphorylation in Leydig cells." *Chemosphere* 185: 237-246.

[33] Feng, Yixing S., et al. 2018. "Mechanism of bisphenol AF-induced progesterone inhibition in human chorionic gonadotrophin-stimulated mouse Leydig tumor cell line (mLTC-1) cells." *Environmental Toxicology* 33.6: 670-678.

[34] Blount, Benjamin C., et al. 2000. "Quantitative detection of eight phthalate metabolites in human urine using HPLC− APCI-MS/MS." *Analytical Chemistry* 72.17: 4127-4134.

[35] Adibi, Jennifer J., et al. 2003. "Prenatal exposures to phthalates among women in New York City and Krakow, Poland." *Environmental health perspectives* 111.14: 1719-1722.

[36] Rudel, Ruthann A., et al. 2003. "Phthalates, alkylphenols, pesticides, polybrominated diphenyl ethers, and other endocrine-disrupting compounds in indoor air and dust." *Environmental science & technology* 37.20: 4543-4553.

[37] Fisher, Michael T., Mitzi Nagarkatti, and Prakash S. Nagarkatti. 2005. "Aryl hydrocarbon receptor-dependent induction of loss of mitochondrial membrane potential in epididydimal spermatozoa by 2, 3, 7, 8-tetrachlorodibenzo-p-dioxin (TCDD)." *Toxicology Letters* 157.2: 99-107.

[38] Chen, Xuemei, et al. 2011. "The combined toxicity of dibutyl phthalate and benzo (a) pyrene on the reproductive system of male Sprague Dawley rats in vivo." *Journal of Hazardous Materials* 186.1: 835-841.

[39] Sedha, Sapna, Sunil Kumar, and Shruti Shukla. 2015. "Role of oxidative stress in male reproductive dysfunctions with reference to phthalate compounds." *Urology Journal* 12.5: 2304-2316.

[40] Li, Linxi, et al. 2016. "Comparison of the effects of dibutyl and monobutyl phthalates on the steroidogenesis of rat immature Leydig cells." *BioMed Research International* 2016: 12.

[41] Zhao, Yan, et al. 2012. "Mono-(2-ethylhexyl) phthalate affects the steroidogenesis in rat Leydig cells through provoking ROS perturbation." *Toxicology in Vitro 26*.6: 950-955.

[42] Zhou, Liang, et al. 2013. "Oxidative stress and phthalate-induced down-regulation of steroidogenesis in MA-10 Leydig cells." *Reproductive Toxicology* 42: 95-101.

[43] Robinson, Brett H. 2009. "E-waste: an assessment of global production and environmental impacts." *Science of the Total Environment* 408.2: 183-191.

[44] McKinney, James D., and Chris L. Waller. 1994. "Polychlorinated biphenyls as hormonally active structural analogues." *Environmental Health Perspectives* 102.3: 290-297.

[45] Schantz, Susan L., John J. Widholm, and Deborah C. Rice. "Effects of PCB exposure on neuropsychological function in children." *Environmental Health Perspectives* 111.3 (2003): 357.

[46] Moysich, Kirsten B., et al. 1998. "Environmental organochlorine exposure and postmenopausal breast cancer risk." *Cancer Epidemiology and Prevention Biomarkers* 7.3: 181-188.

[47] Murugesan, Palaniappan, et al. 2005. "Studies on the protective role of vitamin C and E against polychlorinated biphenyl (Aroclor 1254)—induced oxidative damage in Leydig cells." *Free Radical Research* 39.11: 1259-1272.

[48] Safe, Stephen B., et al. 1985. "PCBs: structure–function relationships and mechanism of action." *Environmental Health Perspectives* 60: 47-56.

[49] Sathish Kumar, T., et al. 2017. "Lactational exposure of polychlorinated biphenyls downregulates critical genes in Leydig cells of F1 male progeny (PND 21)." *Andrologia,* 49.8: e12734.

[50] Thangavelu, Sathish K., et al. 2018. "Lactational exposure of polychlorinated biphenyls impair Leydig cellular steroidogenesis in F1 progeny rats." *Reproductive Toxicology* 75: 73-85.

[51] Aydin, Yasemin, and Melike Erkan. 2017. "The toxic effects of polychlorinated biphenyl (Aroclor 1242) on TM3 Leydig cells." *Toxicology and Industrial Health* 33.8: 636-645.

[52] Rana, S. V. S. 2014. "Perspectives in endocrine toxicity of heavy metals—a review." *Biological Trace Element Research* 160.1: 1-14.

[53] Wirth, Julia J., and Renee S. Mijal. 2010. "Adverse effects of low-level heavy metal exposure on male reproductive function." *Systems Biology in Reproductive Medicine* 56: 147-167.

[54] Ji, Xunmin, et al. 2015. "Cytotoxic mechanism related to dihydrolipoamide dehydrogenase in Leydig cells exposed to heavy metals." *Toxicology* 334: 22-32.

[55] Wen, Luona, et al. 2018. "Cyanidin-3-O-glucoside promotes the biosynthesis of progesterone through the protection of mitochondrial function in Pb-exposed rat leydig cells." *Food and Chemical Toxicology* 112: 427-434.

[56] Li, Xiaojun, et al. 2018. "In utero single low-dose exposure of cadmium induces rat fetal Leydig cell dysfunction." *Chemosphere* 194: 57-66.

[57] Alamdar, Ambreen, et al. 2019. "Enhanced histone H3K9 tri-methylation suppresses steroidogenesis in rat testis chronically exposed to arsenic." *Ecotoxicology and Environmental Safety* 170: 513-520.

[58] Kong, Deying, et al. 2016. "Acetamiprid inhibits testosterone synthesis by affecting the mitochondrial function and cytoplasmic adenosine triphosphate production in rat Leydig cells." *Biology of Reproduction* 96.1: 254-265.

[59] Walsh, L. P., et al. 2000. "Roundup inhibits steroidogenesis by disrupting steroidogenic acute regulatory (StAR) protein expression." *Environmental Health Perspectives* 108.8: 769-776.

[60] Ye, Xiaoqing, et al. 2017. "Pyrethroid Insecticide Cypermethrin Accelerates Pubertal Onset in Male Mice via Disrupting

Hypothalamic–Pituitary–Gonadal Axis." *Environmental Science & Technology* 51.17: 10212-10221.

[61] Zhang, Shu Y., et al. 2007. "Permethrin may disrupt testosterone biosynthesis via mitochondrial membrane damage of Leydig cells in adult male mouse." *Endocrinology* 148.8: 3941-3949.

[62] Eze, Ukpai A., et al. 2019. "In vitro effects of single and binary mixtures of regulated mycotoxins and persistent organochloride pesticides on steroid hormone production in MA-10 Leydig cell line." *Toxicology in Vitro* 60: 272-80.

[63] Liu, Shiwen, et al. 2016. "In utero methoxychlor exposure increases rat fetal Leydig cell number but inhibits its function." *Toxicology* 370: 31-40.

BIOGRAPHICAL SKETCH

Banu Orta Yilmaz

Affiliation: Istanbul University, Faculty of Science.

Education: Biology, PhD.

Business Address: Department of Biology, Faculty of Science, Istanbul University, 34134 Vezneciler, Fatih; Istanbul, Turkey.

Research and Professional Experience: Toxicology, Male Reproductive System, Developmental Biology.

Publications from the Last 3 Years:

Aydin Y., Orta Yilmaz B., Yildizbayrak N., Korkut A., Arabul Kursun M., Irez T., et al., "Evaluation of citrinin-induced toxic effects on mouse Sertoli cells.," *Drug and Chemical Toxicology*, vol.-, pp. 1-7, 2019.

Orta Yilmaz B., Erkan M. B., "Effects of Vitamin C on Antioxidant Systems and Steroidogenic Enzymes in Sodium Fluoride-Exposed Tm4 Sertoli Cells," *Fluoride*, pp. 139-151, 2014.

Orta Yilmaz B., Erkan M. B., "Effects of Vitamin C on Sodium Fluoride-Induced Oxidative Damage in Sertoli Cells," *Fluoride*, pp. 241-251, 2015.

Orta Yilmaz B., Erkan M. B., "The Potential Role of Stem Cell Reprogramming in Antiaging," in: *Molecular Basis and Emerging Strategies for Anti-aging Interventions,* Rizvi S. I., Çakatay U., Eds., Springer, London/Berlin, Singapore, pp. 35-45, 2018.

Orta Yilmaz B., Korkut A., Erkan M. B., "Sodium fluoride disrupts testosterone biosynthesis by affecting the steroidogenic pathway in TM3 Leydig cells," *Chemosphere*, pp. 447-455, 2018.

Orta Yilmaz B., Yildizbayrak N., Aydin Y., Erkan M. B., "Evidence of acrylamide- and glycidamide-induced oxidative stress and apoptosis in Leydig and Sertoli cells.," *Human & Experimental Toxicology*, no. 1, pp. 1-11, 2016.

Orta Yilmaz B., Yıldızbayrak N., Erkan M. B., "Sodium arsenite-induced detriment of cell function in Leydig and Sertoli cells: the potential relation of oxidative damage and antioxidant defense system," *Drug and Chemical Toxicology*, pp. 1-9, 2018.

Perker M. C., Orta Yilmaz B., Yildizbayrak N., Aydin Y., Erkan M. B., "Protective effects of curcumin on biochemical and molecular changes in sodium arsenite-induced oxidative damage in embryonic fibroblast cells.," *Journal of Biochemical and Molecular Toxicology*, vol.-, pp. e22320-e22320, 2019.

In: Leydig Cells
Editor: Bruno Solomon
ISBN: 978-1-53617-282-9

Chapter 4

EFFECT OF ENDOCRINE DISRUPTORS (PCBS/DEHP) ON TESTICULAR LEYDIG CELLS IN ADULT AND F1 MALE OFFSPRING RATS

Arunakaran Jagadeesan, **Murugesan Palaniappan, Elumalai Perumal, Suganya Sekaran and Sathish Kumar Thangavelu***
Department of Endocrinology, Dr. ALM. P.G. Institute of Basic Medical Science, University of Madras, Taramani, Chennai, India

ABSTRACT

Leydig cells are present in the interstitial compartment of the mammalian testis and their main function is to produce testosterone which is essential for spermatogenesis and development of secondary sexual characters. LH is the primary regulator of Leydig cell function. Endocrine disruptors such as polychlorinated biphenyls, diethyl hexyl phthalate which are environmental contaminants cause adverse effects on

---

* Correspondence Author's Email: j_arunakaran@hotmail.com.

Leydig cell structure and function hypertrophy and reduced capacity to produce testosterone. Our earlier studies proved that these disrupters altered the hypothalamic pituitary testicular axis in adult rats. The depletion of Leydig cellular antioxidant enzymes and increase in the levels of reactive oxygen species (ROS) and lipid peroxidation were observed in PCB exposed adult rats. Our studies demonstrated that PCB inhibits testosterone biosynthesis, Leydig cellular LH receptors, steroidogenic enzymes and antioxidant enzymes in adult rats. Lactational exposure of PCBs downregulated critical genes in Leydig cells of F1 male progeny. *In utero* exposure of phthalate downregulates critical genes in Leydig cells of F1 male progeny. Gestational exposure of phthalate induced hypermethylation in SP1 and Sp1 promoter in Leydig cells of F1 offspring rats.

Lycopene is a highly efficient antioxidant and has a singlet-oxygen and free radical scavenging capacity. It has protective effects against testicular toxicity, spermiotoxicity, cardiotoxicity and nephrotoxicity. Our studies demonstrated that lycopene supplementation improved and maintained the normal level of testosterone via improving the expression of StAR protein and steroidogenic enzymes in Leydig cells of PCB exposed rats. Ameliorative effect of α-tocopherol on PCB induced testicular Sertoli cell dysfunction in F1 prepubertal rats was also studied. Extensive studies on Leydig cells are in progress.

INTRODUCTION

In men, the reproductive system has evolved for continues lifelong gamatogenesis, coupled to occasional internal insemination with a high density of sperm. Men produce haploid gametes called sperm essentially in a continuous manner whereas female gamete called the egg or ovum is produced in a discontinuous manner at a rate of 1 egg per month. In adult men the basic roles of gonadal hormones are 1) support of gamatogenesis, spermatogenesis 2) maintenance of male reproductive tract and production of semen and 3) maintenance of secondary sex characteristics and libido. There is no overall cyclicity of this activity in men. The testes reside outside of the abdominal cavity in the scrotum. The location maintains the testicular temperature at about 35°C which is crucial for sperm development. Failure of the testis to descend through the inguinal canal into the scrotum during development results in infertility. The human testis is covered by a connective tissue capsule and is divided into lobules by fibrous septa. Within each lobule are two or four loops of seminiferous tubules. Each loop empties into an anastomosing network of tubules called rate testis. Rete testes is continuous with small ducts called efferent ductless that lead the sperm out of the testis into the head of the epididymis from epididymal to the vas deferens. In the epididymal region the sperm pass from the head, to the body and to the tail of the epididymis and then to the vas deferens. Spermatozoa are stored in the tail of epididymis and the vas deference.

The presence of seminiferous tubules of the testis creates two compartment with in each lobule; an intratubular compartment which is composed of the seminiferous epithelium of the seminiferous tubule and a peritubular compartment which is composed of the seminiferous epithelium of the seminiferous tubule and a peritubular compartment which is composed neurovascular elements, connective tissue cells, immune cells and the interstitial cells of Leydig whole main function is to produce testosterone. The intratubular compartment is lined with seminiferous epithelium composed of two cell types: 1. Sperm cells in various stages of spermatogenesis. 2. The Sertoli cells which is a nurse cell in intimate contact with all sperm cells and which regulates many aspects of spermatogenesis. The Sertoli cells provide instrumental support within epithelium and form adherens-type junctions and gap junctions with all stages of sperm cells. The Sertoli-Sertoli cell occluding junctions divide the seminiferous epithelium into a basal compartment, containing the spermatogonia and early stage primary spermatocytes and an luminal compartment containing late-stage primary spermatocytes and all subsequent stages of sperm cells. The tight junctions form the physical basis for the blood-testis barrier which creates a specialized, immunologically safe microenvironment for developing sperm. The specific functions of Sertoli cells are supportive, exocrine and endocrine. Supportive means maintaining, breaking and re-forming multiple functions with developing sperm, maintaining blood-testis-barrier, phagocytosis, transfer of nutrients, expression of paracrine factors and receptors for sperm derived paracrine factors. Exocrine means production of fluid to more immobile sperm out of testis towards epididymis, production of androgen binding protein and determination of release of spermatozoa (spermiation). Endocrine function means expression of androgen receptor, FSH receptor, production of anti mullerian hormone and aromatization of testosterone to estradiol 17 β (Mruk and cheng, 2004).

The peritubular compartment contains the primary endocrine cells of the testis the Leydig cell. It contains common cell types of loose connective tissue and an extremely rich peritubular capillary network. Leydig cells are steroidogenic stromal cells. These cells synthesis cholesterol *de novo*, as well as acquiring it through low density lipoproteins receptors (LDL receptors) and hight-density lipoproteins receptors (HDL receptor). HDL receptor is also scavenging receptor B1 and store cholesterol as cholesterol esters. Free cholesterol generates by a cholesterol esters hydrolase and transferred to the outer mitochondria membrane and then to the inner mitochondrial membrane in a steroidogenic acute regulatory protein (StAR) dependent manner (Porterfield and white, 2007). Cholesterol is converted to pregnenolone by CYP11A1. Pregnenolone is then processed to progesterone, 17α-hydroxy progesterone and androstenedione by 3β-hydroxy steroid dehydrogenase and CYP17. CYP17 is a functional enzyme with a 17 hydroxylase activity and a 17, 20 lyase activity. CYP17 displays a robust level of both activities in Leydig cell. It expresses a high level of 3β-HSD so that delta 4 pathway is ultimately favored. Leydig cell express a Leydig cell-specific isoform of 17β hydroxysteroid dehydrogenase –type3 where converts androstenedione to testosterone. Testosterone is essential for spermatogenesis and development of secondary sex characters. LH is the primary regulator of Leydig cell function (Bhasin et al., 2003).

Polychlorinated biphenyls are industrial chemicals used in plasticizers, surface coating, links, adhesives, flame retardants, paints days for carbonless duplicating papers. They are heat stable and have been used in dielectric fluids in transformers and capacitors. Their characteristics low solubility, contribute to its ability to bio concentrate which leads to bioaccumulation. PCBs have been regulated as food contaminates and in food syrups. PCBs are distributed throughout the entire ecosystem including soil, air, and water. They are also likely to bioaccumulated in the food chain because of their lipophilicity and therefore belong to a class of environmental chemicals called persistent bioaccumulative toxicants (PBTs) (Fisher, 1999). They are environmental toxicants associated with numerous adverse health effects through widespread bioaccumulation in the biosphere and bio concentration in the food chain (Norstrom et al., 2010). PCB induced toxic manifestations are associated with the production of free radicals. PCB exposure is positively correlated with a decreased IQ stores, impaired learning and memory, decreased neuromuscular function and lower reading comprehension. Developmental exposure to PCBs causes cognitive and psychomotor defects. PCB induced cytotoxicity has been implicated in ROS generation due to the depletion of antioxidants in Leydig cells. PCBs induce cytochrome P450 as a possible source of ROS or alternatively PCB derivatives undergo redox cycling with the formation of ROS like O_2, HO and H_2O_2 thus becoming another source of oxidative stress the effects of PCBs are mediated by the aryl hydrocarbon receptor (Safe, 1995) with high affinity.

Di-(2-ethylhexyl) phthalate (DEHP), a plasticizer commonly used by industry to add flexibility to polyvinyl chloride products, is highly prevalent in the environment. DEHP is present in a wide variety of consumer products, including medical devices such as intravenous tubing, catheters, and dialysis bags. They are not covalently bound with the plastics in which they are mixed and easily leach into the environment (Pak et al., 2006). Phthalates cross from maternal blood into the developing fetus via placental transfer and into neonates via breast milk. These exposures may affect the developing endocrine system, which is essential for diverse biological functions, including sexual development and reproductive functions in adults (Kavlock et al., 2002). Phthalates were also reported to induce adverse reproductive affects in females. Phthalate-induced changes in reactive oxygen species (ROS) have been studied and it causes changes in Leydig cellular steroidogenesis. Peroxisome proliferator activated receptor (PPAR) play a number of important roles in normal physiology. PPARs generally form a heterodimer with the retinoid –x-receptor (RXR) before translocating to the nucleus and binding to the

PPAR on DNA and subsequently activating transcription of specific target genes. The biological action of DEHP is very stimulated to chemicals that are collecting known as peroxisome proliferators. These steroids hormone receptors act as transcription factors upon ligand binding. DEHP may cause epigenetic changes expression (Pogribry et al., 2008). Phthalate exposure impairs insulin receptor and glucose transporter 4 gene expression in L6 myotubes. Phthalate is associated with insulin resistance in adipose tissue of male rats (Rajesh et al., 2013).

Phthalates are synthetic additives, used as plasticizers and are used in a vast range of consumer products including perfume; nail polish, vinyl flooring, toys and medical devices. Phthalates Di (2-ethylhexyl) Phthalates are not covalently bound to PUC, they may be released when children place PUC products in their mouth. DEPH is a lipophilic compound and can be associated in seminal fluid, amniotic fluid, breast milk, saliva and placenta (Mortenson et al., 2005). In human's studies are available assessing the association between phthalate exposure and male reproductive and developmental endpoints such as decreased urogenital distance, hypospadias, cryptorchidism and gynecomastia Phthalates exposure decreased serum hormones such as LH, FSH, estradiol and testosterone and increased SHBG (Jurewicz et al., 2013). Adverse semen functions such as low sperm concentration and motility, sperm malformation and increased DNA damage (Rozati et al., 2002; Hauser et al., 2007). In animal studies phthalate produced a variety of adverse effects including thyroid, liver, and kidney signaling, immune and metabolic homeostasis.

Environmental pollutants such as PCBs, DEHP etc. have caused adverse effects on male reproduction. Exposure to polychlorinated biphenyls has associates with defects in spermatogenesis and reduced weights of testis and accessory sex organs (Murugesan et al., 2005). Rozati et al. (2002) studied that in humans, the levels of PCBs have been inversely correlated to the sperm number and motility. Kim and Ariyatine (2001) found that a continuous exposure of lactating mothers to PCB causes significant effects of Leydig cell structure and function, hypertrophy and reduced capacity to produce testosterone *in vitro* in response to LH stimulation. LH secreted from the anterior pituitary gland regulates testosterone biosynthesis in Leydig cells. LH secretion is regulated by gonadotrophin-releasing hormone released from the hypothalamus. Earlier studies demonstrated that PCB altered the hypothalamic-pituitary-testicular axis and disrupt Sertoli and Leydig cellular functions in adult rats (Murugesan et al., 2005a,b; Muthuvel et al., 2006; Senthilkumar et al., 2004; Krishnamoorthy et al., 2005). The cultured Leydig cells from adult rats exposed to PCBs resulted in lowered synthesis of testosterone (Murugesan et al., 2007; 2008).

Our earlier studies demonstrated that the steroidogenic enzymes such as cytochrome P450 scc, 3β and 17β hydroxysteroid dehydrogenase activities are diminished in PCB exposed adult rat Leydig cells *in vivo* (Murugesan et al., 2005a). In addition, the depletion of Leydig cellular antioxidant enzymes and increase in the levels of reactive oxygen species (ROS) and lipid peroxidase (LPO) were observed in PCB exposed adult rats (Murugesan et al., 2007). Lycopene, a naturally occurring carotenoid which has been attracted considerable attention as a potential chemopreventive agent and antioxidant. Recent studies have reported that lycopene is a highly efficient antioxidant and has a singlet-oxygen and free radical scavenging capacity (Cohen, 2002; Tapiero et al., 2004). It has protective effects against testicular toxicity, spermatotoxicity, cardiotoxicity and nephrotoxicity (Atessahin et al., 2005; 2006; Karahan et al., 2005; Yilmaz et al., 2006; Turk et al., 2007). Our earlier study, Elumalai et al., (2009) demonstrated that lycopene treatment maintained normal level of testosterone with improving the expression of StAR protein and this may be due to normal LH and LH receptor availability and steroidogenic enzyme activity.

Environmental contaminants are endocrine disruptors. They interfere with or mimic estrogen and androgen action on Leydig cells (Kwan et al., 1993). The anti-inflammatory drugs such as indomethacin and chloroquine have been used a tools demonstrate the role of prostaglandins in steroidogenesis (Sawada et al., 1994). Our earlier studies have been used as tools demonstrate that PCB is antiestrogen and that this is achieved by the downregulation of the estrogen receptor (Safe and Krishnan, 1995). PCB binds to the arylhydrocarbon receptor, a member of the nuclear hormone receptor superfamily and acts as a transcription factor inhibiting the expression of some estrogen responsive genes while promoting others (Safe, 1994). Apart from AhR, it can also act at estrogen receptor, androgen receptor and thyroid receptor. Our earlier studies we proved that PCB caused neuronal damage and alters the antioxidant system and is also induces NMDA receptor mediated glutamate excitotoxicity in adult male rats (Venkataraman et al., 2010; Selvakumar et al., 2013; Bavithra et al., 2015). The reproductive toxicology of PCBs have been extensively studied that exposure of PCBs induced hypothyroidism, reduced sperm count, motility, altered menstrual cycle, reduced conception rate and it also causes spontaneous abortion (Krishnamoorthy et al., 2013; Xiao et al., 2011; Arunakaran, 2016). Our previous studies proved that exposure of PCBs downregulates critical genes in Leydig cells of adult as well as F1 male progeny (Murugesan et al., 2005a,b; Elumalai et al., 2009; Sathish kumar et al., 2017).

In addition to PCB we studied that lactational exposure of phthalate causes long term disruption in testicular architecture by altering tight junctional and apoptotic protein expression in Sertoli cells of first flial generation pubertal wistar rats (Sekaran et al., 2015). *In utero* exposure to Phthalates downregulates critical genes in Leydig cells of F1 male progeny (Sekaran and Jagadeesan, 2015). Gestational exposure of DEPH induced hypomethylation in SF1 and SP1 promotor in Leydig cell thereby it affects the Leydig cell function in F1 offspring. Methylation was observed in the promoter of SF1. DEHP transfer can occur from mother to offspring and information concerning the effects of neonatal exposure via milk in the regulation of epigenetic mechanisms on testis of F1 offspring was observed (Sekeran and Jagadeesan, 2015).

Our recent study found that exposure to PCBs decreased the body weight, testis weight and urogenital distance index in F1 progeny rats. PCBs exposure reduced the serum levels of LH, testosterone and estradiol (Sathish kumar et al., 2017). Interestingly PCBs caused a decrease in Leydig cell population along with the decreased activities of steroidogenic enzymes 3β and 17βHSD. Additionally we observed a significant decrease in LHR, SR-B1, StAR protein, CYP11a1, 3β HSD, CYP17a1, 17β HSD, 5 alpha reductase, Cyp 19 a1 and AR gene expression in the Leydig cells of progeny rats (Thangavel et al., 2018). Therefore, in essence we proved that PCBs exposure alters Leydig cellular steroidogenesis in the F1 progeny rats. Recently Arunakaran, (2018) reviewed endocrine disrupting chemicals on female reproduction. Alpha tocopherol has ameliorative role against PCBs induced testicular dysfunction in F1 progeny rats (Elyapillai et al., 2017). PCBs decreased the protein levels of FSHR, AR ABP, ERα and β, transferrin, claudin II, occuludin, E-cahderin, connexin 43, C-fos, C-Jun, SF1, USF 1 and 2 whereas inhibin β protein level was found to be increased in Sertoli cells of F1 progeny rats (Elayapillai et al., 2017). Endocrine disruptors can cause the epigenetic transgenerational effects in F1 through F4 generation.

To conclude, our studies demonstrated that lactational exposure of PCBs/DEHP affects the testicular architecture, Leydig and Sertoli cellular functions by disturbing the level of functional regulators, transcription factors regulating their function through epigenetic mediated mechanism in Leydig and Sertoli cells of F1 pubertal rats. The extensive studies on Leydig cells are in progress.

Acknowledgments

The financial assistance from the Department of Science and Technology (DST), Indian council of Medical Research, Government of India, New Delhi and University of Madras to Dr J. Arunakaran is greatly acknowledged.

Conflict of Interest

The authors declared no potential conflict of interest with respect to authorship and or publication of this article.

References

Arunakaran J (2016). Effects of PCB and phthalate on male reproductive system in adult and F11 male offspring – A review. *ISSRF Newsletter* issue 18, ISSN 2395 – 2806 pp 48-51.

Arunakaran J (2018). Endocrine disrupting chemicals on female reproduction. *Obstect. Gycecol Int J.* 9(6): 510-512.

Ateşşahin A, Karahan I, Türk G, Gür S, Yilmaz S, Ceribaşi AO (2006). Protective role of lycopene on cisplatin-induced changes in sperm characteristics, testicular damage and oxidative stress in rats. *Reprod Toxicol.* 21(1): 42-7.

Ateşşahin A, Türk G, Karahan I, Yilmaz S, Ceribaşi AO, Bulmuş O (2006) Lycopene prevents adriamycin-induced testicular toxicity in rats. *Fertil Steril.* 85 Suppl 1:1216-22.

Bavithra S, Priya ES, Selvakumar K, et al. (2015) Effect of melatonin on glutamate: BDNF signaling in the cerebral cortex of polychlorinated biphenyls (PCBs)—exposed adult male rats. *Neurochem Res.* 40(9): 1858–1869.

Bhasin S, Enzlin P, Coviello A, Basson R (2007). Sexual dysfunction in men and women with endocrine disorders. *Lancet.* 369(9561): 597-611. Review.

Cohen LA (2002). A review of animal model studies of tomato carotenoids, lycopene, and cancer chemoprevention. *Exp Biol Med (Maywood).* 227(10): 864-8. Review.

Elayapillai SP, Teekaraman D, Paulraj RS, et al. (2017). Ameliorative effect of α-tocopherol on polychlorinated biphenyl (PCBs) induced testicular Sertoli cell dysfunction in F1 prepuberal rats. *Exp Toxicol Pathol.* 69(8): 681–694.

Elumalai P, Krishnamoorthy G, Selvakumar K, et al. (2009). Studies on the protective role of lycopene against polychlorinated biphenyls (Aroclor 1254)-induced changes in StAR protein and cytochrome P450 scc enzyme expression on Leydig cells of adult rats. *Reprod Toxicol.* 27(1): 41–45.

Fisher B. E (1999), "Most unwanted," *Environmental Health Perspectives,* vol. 107, no. 1, pp. A18–A23.

Hauser R, Meeker JD, Singh NP, Silva MJ, Ryan L, Duty S, Calafat AM (2007). DNA damage in human sperm is related to urinary levels of phthalate monoester and oxidative metabolites. *Hum Reprod.* 22(3): 688-95.

Jurewicz J, Radwan M, Sobala W, Ligocka D, Radwan P, Bochenek M, Hawuła W, Jakubowski L, Hanke W (2013). Human urinary phthalate metabolites level and main semen parameters, sperm chromatin structure, sperm aneuploidy and reproductive hormones. *Reprod Toxicol.* 42: 232-41.

Karahan I, Ateşşahin A, Yilmaz S, Ceribaşi AO, Sakin F (2005). Protective effect of lycopene on gentamicin-induced oxidative stress and nephrotoxicity in rats. *Toxicology.* 215(3): 198-204.

Kavlock R, Boekelheide K, Chapin R, Cunningham M et al. (2002). Center for the evaluation of risks to human reproduction; Phthalates expert panel report on the reproductive and developmental toxicity of di (2-ethylhexyl) phthalate. *Reprod Toxicol.* 16: 529–653.

Kim IS, Ariyaratne HB, Chamindrani Mendis-Handagama SM (2001). Effects of continuous and intermittent exposure of lactating mothers to aroclor 1242 on testicular steroidogenic function in the adult male offspring. *Tissue Cell.* 33(2):169-77.

Krishnamoorthy G, Murugesan P, Muthuvel R, Gunadharini DN, Vijayababu MR, Arunkumar A, Venkataraman P, Aruldhas MM, Arunakaran J (2005). Effect of Aroclor 1254 on Sertoli cellular antioxidant system, androgen binding protein and lactate in adult rat in vitro. *Toxicology.* 212(2-3): 195-205.

Krishnamoorthy G, Selvakumar K, Venkataraman P, et al. (2013). Lycopene supplementation prevents reactive oxygen species mediated apoptosis in Sertoli cells of adult albino rats exposed to polychlorinated biphenyls. *Interdiscip Toxicol.* 6(2):83–92.

Kwan TK, Foong SL, Lim YT, Gower DB (1993). Effects of some non-steroidal anti-inflammatory drugs (NSAIDS) on steroidogenesis in rat and porcine testis. *Biochem Mol Biol Int.* 31(4): 733-43.

Mortensen GK, Main KM, Andersson AM, Leffers H, Skakkebaek NE (2005). Determination of phthalate monoesters in human milk, consumer milk, and infant formula by tandem mass spectrometry (LC-MS-MS). *Anal Bioanal Chem.* 382(4): 1084-92.

Mruk DD, Cheng CY (2004). Sertoli-Sertoli and Sertoli-germ cell interactions and their significance in germ cell movement in the seminiferous epithelium during spermatogenesis. *Endocr Rev.* 25(5):747-806. Review.

Murugesan P, Balaganesh M, Balasubramanian K, Arunakaran J (2007). Effects of polychlorinated biphenyl (Aroclor 1254) on steroidogenesis and antioxidant system in cultured adult rat Leydig cells. *J Endocrinol.* 192(2):325-38.

Murugesan P, Kanagaraj P, Yuvaraj S, Balasubramanian K, Aruldhas MM, Arunakaran J (2005). The inhibitory effects of polychlorinated biphenyl Aroclor 1254 on Leydig cell LH receptors, steroidogenic enzymes and antioxidant enzymes in adult rats. *Reprod Toxicol.* 20(1):117-26.

Murugesan P, Muthusamy T, Balasubramanian K, Arunakaran J (2008). Polychlorinated biphenyl (Aroclor 1254) inhibits testosterone biosynthesis and antioxidant enzymes in cultured rat Leydig cells. *Reprod Toxicol.* 25(4):447-54.

Murugesan P, Senthilkumar J, Balasubramanian K, et al. (2005). Impact of polychlorinated biphenyl Aroclor 1254 on testicular antioxidant system in adult rats. *Hum Exp Toxicol.* 24(2):61–66.

Muthuvel R, Venkataraman P, Krishnamoorthy G, et al. (2006). Antioxidant effect of ascorbic acid on PCB (Aroclor 1254) induced oxidative stress in hypothalamus of albino rats. *Clin Chim Acta.* 365(1–2):297–303.

Norström K, Czub G, McLachlan MS, Hu D, Thorne PS, Hornbuckle KC (2010). External exposure and bioaccumulation of PCBs in humans living in a contaminated urban environment. *Environ Int.* 36(8): 855-61.

Pak VM, Briscoe V, McCauley LA (2006). How to reduce DEHP in your NICU: a plan of simple steps to promote change. *Neonatal Netw.* 25: 447- 449.

Pogribny IP, Tryndyak VP, Boureiko A, Melnyk S, Bagnyukova TV, Montgomery B, Rusyn I (2008). Mechanisms of peroxisome proliferator-induced DNA hypomethylation in rat liver. *Mutat Res.* 644(1-2):17-23.

Rajesh P, Sathish S, Srinivasan C, Selvaraj J, Balasubramanian K (2013). Phthalate is ssociated with insulin resistance in adipose tissue of male rat: role of antioxidant vitamins. *J Cell Biochem.* 114(3):558-69.

Rozati R, Reddy PP, Reddanna P, Mujtaba R (2002). Role of environmental estrogens in the deterioration of male factor fertility. *Fertil Steril.* 78(6):1187-94.

Safe SH and Krishen V (1995). Cellular and molecular biology of aryhydrocarbon (Ah) receptor mediated gene expression. *Arch Toxicol Suppl* 17: 99 – 115.

Safe SH (1994). Polychlorinated biphenyls (PCBs): Environmental impact, biochemical and toxic responses, and implications for risk assessment. *Crit Rev Toxicol.* 24(2): 87-149. Review.

Sathish Kumar T, Sugantha Priya E, Raja Singh P, et al. (2017). Lactational exposure of polychlorinated biphenyls downregulates critical genes in Leydig cells of F1 male progeny (pnd21). *Andrologia.* 49(8).

Sawada T, Asada M, Mori J (1994). Effects of single and repeated administration of prostaglandin F2 alpha on secretion of testosterone by male rats. *Prostaglandins.* 47(5): 345-52.

Sekaran S, Balaganapathy P, Parsanathan R, et al. (2015). Lactational exposure of phthalate causes long-term disruption in testicular architecture by ltering tight junctional and apoptotic protein expression in Sertolicells of first filial generation pubertal Wistar rats. *Hum Exp Toxicol.* 34(6): 575–590.

Sekaran S, Jagadeesan A (2015). In utero exposure to phthalate downregulates critical genes in Leydig cells of F1 male progeny. *J Cell Biochem.* 116(7): 1466–1477.

Selvakumar K, Bavithra S, Ganesh L, et al. (2013). Polychlorinated biphenyls induced oxidative stress mediated neurodegeneration in hippocampus and behavioral changes of adult rats: anxiolytic-like effects of quercetin. *Toxicol Lett.* 222(1): 45–54.

Senthilkumar J, Banudevi S, Sharmila M, et al. (2004). Effects of Vitamin C and E on PCB (Aroclor 1254) induced oxidative stress, androgen binding protein and lactate in rat Sertoli cells. *Reprod Toxicol.* 19(2): 201–208.

Tapiero H, Townsend DM, Tew KD (2004). The role of carotenoids in the prevention of human pathologies. *Biomed Pharmacother.* 58(2):100-10. Review.

Thangavelu SK, Elaiyapillai SP, Ramachandran I, et al. (2017). Lactational exposure of polychlorinated biphenyls impairs Leydig cellula steroidogenesis in F 1 progeny rats. *Reproductive Toxicology.* 75:73–85.

Türk G, Ateşşahin A, Sönmez M, Yüce A, Ceribaşi AO (2007). Lycopene protects against cyclosporine A-induced testicular toxicity in rats. *Theriogenology*. 67(4): 778-85.

Venkataraman P, Selvakumar K, Krishnamoorthy G, Muthusami S, Rameshkumar R, Prakash S, Arunakaran J (2010). Effect of melatonin on PCB (Aroclor 1254) induced neuronal damage and changes in Cu/Zn superoxide dismutase and glutathione peroxidase-4 mRNA expression in cerebral cortex, cerebellum and hippocampus of adult rats. *Neurosci Res.* 66(2):189-97.

Xiao W, Zhang J, Liang J, Zhu H, Zhou Z, Wu Q (2011). Adverse effects of neonatal exposure to 3,3',4,4',5,5'-hexachlorobiphenyl on hormone levels and testicular function in male Sprague-Dawley rats. *Environ Toxicol.* 26(6):657-68.

Yilmaz S, Atessahin A, Sahna E, Karahan I, Ozer S (2006). Protective effect of lycopene on adriamycin-induced cardiotoxicity and nephrotoxicity. *Toxicology.* 218(2-3):164-71.

Bibliography

Cellular and molecular regulation of testicular cells

LCCN	95046670
Type of material	Book
Main title	Cellular and molecular regulation of testicular cells / Claude Desjardins, editor.
Published/Created	New York: Springer, c1996.
Description	xviii, 371 p.: ill.; 24 cm.
ISBN	0387946489 (New York: hardcover: acid-free paper)
LC classification	QP255 .C457 1996
Related names	Desjardins, Claude, 1938- Serono Symposia, USA. Testis Workshop on Cellular and Molecular Regulation of Testicular Cells (13th: 1995: Raleigh, N.C.)
Subjects	Testis--Physiology--Congresses. Testis--Molecular aspects--Congresses. Spermatogenesis--Congresses. Testis--Cytology--Congresses. Testis--cytology--congresses.

	Spermatogenesis--physiology--congresses.
	Leydig Cells--cytology--congresses.
	Steroids--biosynthesis--congresses.
Notes	"Serono Symposia USA."
	"Proceedings of the XIIIth Testis Workshop on Cellular and Molecular Regulation of Testicular Cells, sponsored by Serono Symposia USA, Inc., held March 30 to April 1, 1995, in Raleigh, North Carolina"--T.p. verso.
	Includes bibliographical references and indexes.

Testosterone: from basic research to clinical applications

LCCN	2013948370
Type of material	Book
Personal name	Smith, Lee B., author.
Main title	Testosterone: from basic research to clinical applications / Lee B. Smith, Rod T. Mitchell, Iain J. McEwan.
Published/Produced	New York: Springer, [2013]
Description	xi, 73 pages: illustrations (chiefly color); 24 cm.
ISBN	9781461489771 (pbk: alk. paper)
	1461489776 (pbk: alk. paper)
LC classification	QP572.T4 S65 2013
Related names	Mitchell, Rod T., author.
	McEwan, I. J., author.
Subjects	Testosterone.
	Testosterone--physiology.
	Androgens--therapeutic use.
	Leydig Cells--metabolism.
	Receptors, Androgen--physiology.
	Testosterone--therapeutic use.
Notes	Includes bibliographical references.
Series	SpringerBriefs in reproductive biology, 2194-4253

SpringerBriefs in reproductive biology. 2194-4253

Testosterone research trends

LCCN	2007000099
Type of material	Book
Main title	Testosterone research trends / L.I. Ardis, editor.
Published/Created	New York: Nova Biomedical Books, c2007.
Description	xiv, 304 p.: ill. (some col.); 27 cm.
Links	Table of contents only http://www.loc.gov/catdir/toc/ecip077/2007000099.html
ISBN	9781600215506 (hardcover) 1600215505 (hardcover)
LC classification	RM296.5.T47 T487 2007
Related names	Ardis, L. I.
Contents	Commentary: Testosterone depletion and cognitive impairment in aging men: a possible relationship between testosterone and Alzheimer's disease? / Hélène Forget, Martine Simard and Séverine Hervouet -- Arachidonic acid and cAMP co-regulation of StAR gene expression and testosterone biosynthesis in aging Leydig cells / XingJia Wang -- Localized modulation of testosterone action: function of steroid receptor coactivators in the brain / Thierry D. Charlier and Jacques Balthazart -- Testosterone and the nigrostriatal dopaminergic system / Dean E. Dluzen -- Testosterone and the brain: implications for cognition, biological rhythms and aging / Ilia N. Karatsoreos, Mary Vernov and Russell D. Romeo -- Androgens and breast cancer: the hyperandrogenic theory of breast cancer: past, present, and future / Giorgio

	Secreto ... [et al.] -- Effects of testosterone on skeletal muscle: the role of exercise / Antonios Matsakas and Michalis G. Nikolaidis -- Testosterone in ischemic stroke and neuronal repair / Yi Pan and Aninda B. Acharya -- Testosterone in kidney ischemia/reperfusion injury / Kwon Moo Park -- Intercollegiate athletics: competition increases saliva testosterone in women soccer, volleyball, and softball players / David A. Edwards ... [et al.] -- Testosterone changes in romantic love / Donatella Marazziti, Mario Catena and Domenico Canale -- Testosterone and its role in inter-individual behavior in a socially living rodent species / E. Scheibler and R. Weinandy -- Testosterone and parasite-mediated sexual selection in a tetraonid bird / François Mougeot and Steve Redpath -- Testosterone degradation by bacteria / Masae Horinouchi, Toshiaki Hayashi and Toshiaki Kudo -- Effects of testosterone in development of electroencephalography in male humans and primates / Adrián Poblano, Carmina Arteaga and Braulio Hernández.
Subjects	Testosterone. Testosterone--Physiology. Testosterone--physiology. Androgens--physiology.
Notes	Includes bibliographical references and index.

The Cell biology of the testis

LCCN	82007985
Type of material	Book
Main title	The Cell biology of the testis / edited by C.

Wayne Bardin and Richard J. Sherins.

Published/Created New York, N.Y.: New York Academy of Sciences, 1982.

Description xii, 543 p.: ill.; 23 cm.

ISBN 0897661826 (pbk.): 0897661818 (hard)

LC classification Q11 .N5 vol. 383 QP255

Related names Bardin, C. Wayne, 1934-
Sherins, Richard J., 1937-
New York Academy of Sciences.
Conference on the Cell Biology of the Testis (1981: New York, N.Y.)

Subjects Testis--Congresses.
Cytology--Congresses.
Testis--Cytology--Congresses.
Leydig cells--Congresses.
Sertoli cells--Congresses.

Notes Proceedings of the Conference on the Cell Biology of the Testis sponsored by the New York Academy of Sciences held in New York City, Apr. 13-15, 1981.
Includes bibliographical references and indexes.

Series Annals of the New York Academy of Sciences; v. 383

The neuroendocrine Leydig cells and their stem cell progenitors, the pericytes

LCCN 2009926054

Type of material Book

Main title The neuroendocrine Leydig cells and their stem cell progenitors, the pericytes / Michail S. Davidoff ... [et al.].

Published/Created Dordrecht; New York: Springer, c2009.

Description ix, 112 p.: ill. (some col.); 24 cm.

ISBN	9783642005121 (pbk.: alk. paper) 3642005128 (pbk.: alk. paper)
LC classification	QP255 .N48 2009
Related names	Davidov, Mikhail S.
Subjects	Leydig cells. Paraneurons. Stem cells.
Notes	Includes bibliographical references (p. [109]-149) and index.
Series	Advances in anatomy, embryology, and cell biology, 0301-5556; 205 Advances in anatomy, embryology, and cell biology; 205.

INDEX

C

D

E

P

R

S

T

U

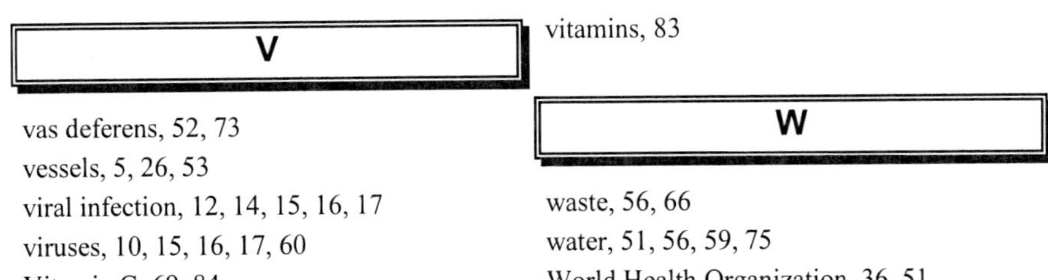
V
W